AF576062

Forest Biodiversity, Conservation and Sustainability

Forest Biodiversity, Conservation and Sustainability

Editor

Petros Ganatsas

MDPI • Basel • Beijing • Wuhan • Barcelona • Belgrade • Manchester • Tokyo • Cluj • Tianjin

Editor
Petros Ganatsas
Laboratory of Silviculture,
Department of Forestry and
Natural Environment,
Faculty of Geotechnical Sciences,
Aristotle University of Thessaloniki
Greece

Editorial Office
MDPI
St. Alban-Anlage 66
4052 Basel, Switzerland

This is a reprint of articles from the Special Issue published online in the open access journal *Sustainability* (ISSN 2071-1050) (available at: https://www.mdpi.com/journal/sustainability/special_issues/Forest_Biodiversity).

For citation purposes, cite each article independently as indicated on the article page online and as indicated below:

LastName, A.A.; LastName, B.B.; LastName, C.C. Article Title. *Journal Name* **Year**, *Volume Number*, Page Range.

ISBN 978-3-0365-0328-8 (Hbk)
ISBN 978-3-0365-0329-5 (PDF)

Cover image courtesy of Petros Ganatsas.

Contents

About the Editor

Petros Ganatsas is a Professor of Forest Ecology and Silviculture in the Department of Forestry and Natural Environment at the Aristotle University of Thessaloniki, Greece. He graduated in Forestry and Natural Environment and obtained his PhD from the School of Agriculture, Forestry and Natural Environment at the Aristotle University of Thessaloniki. His research interests are focused on forest ecosystems structure, functions, and services; biodiversity conservation; ecology; reforestations; climate change and ecosystem management; forest ecology; forest conservation; silvicultural systems; ecosystem ecology; forest habitats; and sustainable management of forest ecosystems. He has published a great number (over 120) of research papers in several national and international peer-reviewed journals while running a wide range of research projects.

Article

Mitigating the Effects of Climate Change through Harvesting and Planting in Boreal Forests of Northeastern China

Xu Luo [1,*], Hong S. He [2], Yu Liang [3], Jacob S. Fraser [2] and Jialin Li [1]

1. Department of Geography & Spatial Information Techniques, Ningbo University, Ningbo 315211, China; nbnj2001@163.com
2. School of Natural Resources, University of Missouri, Columbia, MO 65211, USA; heh@missouri.edu (H.S.H.); fraserjs@missouri.edu (J.S.F.)
3. Institute of Applied Ecology, Chinese Academy of Science, Shenyang 110016, China; liangyu@iae.ac.cn

* Correspondence: luoxu99@hotmail.com; Tel.: +86-574-8760-2769

Received: 11 July 2018; Accepted: 26 September 2018; Published: 1 October 2018

Abstract: The ecological resilience of boreal forests is an important element of measuring forest ecosystem capacity recovered from a disturbance, and is sensitive to broad-scale factors (e.g., climate change, fire disturbance and human related impacts). Therefore, quantifying the effects of these factors is increasingly important for forest ecosystem management. In this study, we investigated the impacts of climate change, climate-induced fire regimes, and forest management schemes on forest ecological resilience using a forest landscape model in the boreal forests of the Great Xing'an Mountains, Northeastern China. First, we simulated the effects of the three studied variables on forest aboveground biomass, growing space occupied, age cohort structure, and the proportion of mid and late-seral species indicators by using the LANDIS PRO model. Second, we calculated ecological resilience based on these four selected indicators. We designed five simulated scenarios: Current fire only scenario, increased fire occurrence only scenario, climate change only scenario, climate-induced fire regime scenario, and climate-fire-management scenario. We analyzed ecological resilience over the five scenarios from 2000 to 2300. The results indicated that the initialized stand density and basal area information from the year 2000 adequately represented the real forest landscape of that year, and no significant difference was found between the simulated landscape of year 2010 and the forest inventory data of that year at the landscape scale. The simulated fire disturbance results were consistent with field inventory data in burned areas. Compared to the current fire regime scenario, forests where fire occurrence increased by 30% had an increase in ecological resilience of 12.4–43.2% at the landscape scale, whereas increasing fire occurrence by 200% would decrease the ecological resilience by 2.5–34.3% in all simulated periods. Under the low climate-induced fire regime scenario, the ecological resilience was 12.3–26.7% higher than that in the reference scenario across all simulated periods. Under the high climate-induced fire regime scenario, the ecological resilience decreased significantly by 30.3% and 53.1% in the short- and medium-terms at landscape scale, while increasing slightly by 3.8% in the long-term period compared to the reference scenario. Compared to no forest management scenario, ecological resilience was decreased by 5.8–32.4% under all harvesting and planting strategies for the low climate-induced fire regime scenario, and only the medium and high planting intensity scenarios visibly increased the ecological resilience (1.7–15.8%) under the high climate-induced fire regime scenario at the landscape scale. Results from our research provided insight into the future forest management and have implications for improving boreal forest sustainability.

Keywords: Ecological resilience; fire disturbance; forest landscape; The Great Xing'an Mountains; LANDIS

1. Introduction

Boreal forests are the northern-most forested biomes, and are expected to be sensitive to climate change and forest fire disturbances [1]. By 2100, the climate in Northern high latitudes (including North America and Eurasia) is expected to increase by 1.4 to 5.8 °C, about 5 times greater than the global mean temperature increase [2,3]. Recent studies have projected that climate change will affect the species distribution and ecosystem functions (e.g., carbon fixation) within boreal forests [4]. For instance, climate change can alter interspecific competition and tree species migration [5], which could further affect forest composition and distribution [6]. Species response to climate change varied among and within populations of different trees in the boreal forests [7]. Changes in climatic conditions (e.g., precipitation and temperature) can have direct influences on the metabolic processes, growth rates, establishment abilities, and competitive ability of trees, thus affecting overall biomass accumulation in forests [5,8]. These resulting changes in forest structure and function are expected to affect the recovery ability (ecological resilience) of boreal forests at landscape scales [9]. Additionally, climate change indirectly impacts forest traits (e.g., forest structure, composition, and ecological resilience) through its effects on fires regimes [10]. In boreal forests, climate-induced fires are frequent and widespread, and become a major factor that can distinctly affect forest successional dynamics, composition, and structure [11,12]. Johnstone et al. [11] showed that forest fires with increased severity may promote shifts from coniferous forest to deciduous-dominated forests, and substantially change landscape dynamics and ecosystem services in boreal forests. More previous studies indicated that fires have been projected to occur more frequently, burn greater areas, and have higher intensities under altered climatic conditions [13,14]. As a result of climate change altered fire regimes, forest composition and biomass dynamics, and thus ecological resilience is expected to shift [15]. Despite the growing evidence that climate change and shifting fire regimes will alter the composition, structure and biomass of boreal forests, quantification of how these two factors will impact forest ecological resilience is still poorly known.

Ecological resilience is characterized as the capacity of a forest ecosystem to recover from disturbance and maintain a stable state, supporting the recovery of structure, composition, and function equivalent to pre-disturbance states [16]. Boreal forests were remarkably resilient to disturbances, and forest species were adapted to the current disturbance regimes with long term effects [17]. For the existence of forest ecological resilience, boreal forest ecosystems thus have the capacity to absorb a spectrum of perturbations (e.g., climate change and forest fires) and to sustain its structure and function, and to maintain the forest ecosystem in a relatively stability domain [16,17]. Climate change in the past century has caused more frequent extreme climate events, such as higher temperatures, severe, and extensive droughts [15], and also has altered forest fires regimes to varying degrees [12,18]. These changing factors (e.g., climate change and climate-induced fires) will exacerbate the loss of ecological resilience in boreal forest ecosystems under long-term exposure [15,19], and may cause a catastrophic shift in forest ecosystems that is difficult to reverse, thus posing a very serious threat to regional ecological security and forest service [20]. Therefore, understanding and quantifying ecological resilience is increasingly important for forest ecosystem management, and provides a quantitative basis for exploring the issue of maintaining and improving ecological resilience.

Harvesting and planting are major anthropogenic disturbances to boreal forests. Boreal harvesting and planting alters forest composition and structure, aboveground biomass accumulation, and ecological resilience from stand to landscape scales [21], and these effects could be aggregated under future changed climate conditions [22,23]. He et al. [22] evaluated species response to harvesting and climate-induced fire in Northern Wisconsin boreal forests, and showed that increased fire frequency can significantly alter the distribution of shade tolerant species, and indicated that harvesting accelerated the decline of Northern hardwood and boreal tree species. Gustafson et al. [23] estimated the climate effects on forest composition in the South-Central Siberian region, and indicated that the direct effects of climate change were not as important as the timber harvesting effects on local virgin forests. However, there are fewer studies exploring the effects of forest management schemes (harvest and planting strategies) on the ecological resilience of boreal forests.

Climate change, fire disturbance, harvesting, and planting occur at large spatio-temporal scales, which makes evaluating their effects on ecological resilience using traditional observation experimental studies challenging [24,25]. Forest landscape models (FLMs) provide a proper scientific approach for studying these issues [26]. With FLMs, we can conduct large-scale studies in which critical model parameters could be changed to explore the complex interactive effects of these extra factors on ecological resilience [27].

The objective of this research was to investigate effects of climate change, climate-induced fire regimes, and future possible forest management schemes on the ecological resilience of boreal forests in Northeastern China. Specifically, we quantified (1) individual effects and (2) interactive effects of climate change and climate-induced fire disturbance on boreal forest ecological resilience, and (3) evaluated whether future possible forest management schemes could mitigate the effects of climate change and climate-induced fires on ecological resilience.

2. Materials and Methods

2.1. Study Area

The study area is located in the Great Xing'an Mountains, which covers nearly 2.7 million ha (Figure 1, 51°35′ to 53°25′ N and 122°25′ to 125°35′ E). The climate conditions are characterized by terrestrial monsoons with long winters and short summers, and the mean monthly temperatures range from −28 to 20 °C in January and July, respectively. Precipitation mainly falls in the summer, and the mean annual value is 428 mm. The elevation ranges from 173 m to 1511 m across the landscape, and the region is covered by brown coniferous forest soils. The dominant vegetation in this area is larch (*Larix gmelinii*) forests. White birch (*Betula platyphylla*) and aspen (*Populus davidiana*) are the major broad-leaved species in this region. In addition to larch and white birch, Mongolian Scots pine (*Pinus sylvestris* var. *mongolica*), and Korean spruce (*Picea koraiensis*) are also widely distributed. Dwarf pine (*Pinus pumila*) has small species communities which can be found in high latitude regions.

Forest fire is a major disturbance in the Great Xing'an Mountains. Based on the Chinese Federal Forest Service data (website: http://www.cfsdc.org), forest fires burned 519,144 ha of the landscape during the period of 1965 to 2005 in this region. For half a century, extreme fire suppression policies have changed fire regimes in this area profoundly. Previous studies indicated that fire regimes have changed from frequent and lower intensity fires to more infrequent and high intensity fires [28]. Extensive harvesting events have affected the forest structure, composition, and natural regeneration significantly in this region. According to the forest inventory data and our field investigation, coniferous dominated forests have shifted from late-seral to mid-seral stages over the landscape. Planting occurs rarely in our study area and overall has minimal effect compared to fire and harvesting. Under the long-term effects of these two typical disturbances (fire and harvesting), the boreal forests of our study region have become more fragmented, simplified, and less resilient [29].

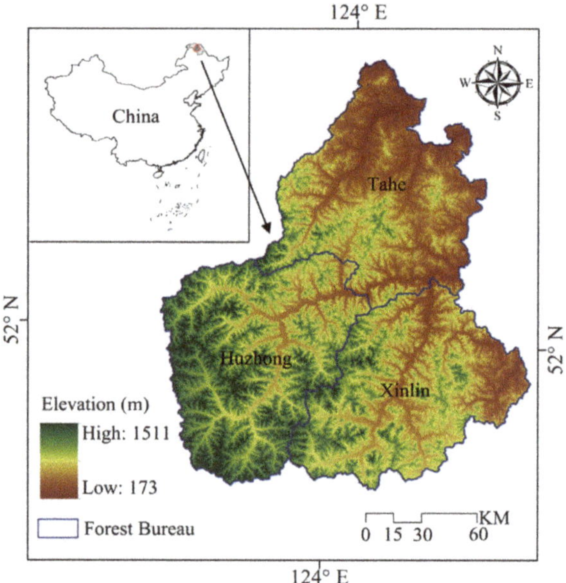

Figure 1. The location of our study area. The insert map shows the location of our study region in Northeastern China.

2.2. The Indicators of Ecological Resilience

The goal of our study was to evaluate the effects of three variables (climate change, climate-induced fire regimes and future possible forest management schemes) on ecological resilience of the study region. Firstly, by using LANDIS PRO model, we simulated the effects of climate variations, fire regimes, and forest management schemes on these four indicators: Aoveground biomass, growing space occupied, age cohort structure, and proportion of mid and late-seral species, which can be obtained directly by LANDIS PRO model outputs. Secondly, we estimated ecological resilience through these four indicators (weighted by a function of the variance). Thus, we can investigate the effects of climate variation, fire regimes, and forest management schemes on boreal ecological resilience in our study area.

Previous studies indicated that the ecological resilience can be quantified through boreal structure, composition, and functioning [15,30]. Seidl et al. [9] and Van Mantgem et al. [31] suggested that the total C storage, the rumple index, the presence of late-seral species and the proportion of older age cohorts can be served as the indicators of ecological resilience. Based on the two previous studies and the current status of our study area, we selected aboveground biomass, growing space occupied, age cohort structure and proportion of mid- and late-seral species as the indicators of ecological resilience (Table 1). The specific contents of these selected indicators are as follows: With regard to boreal forest functioning, we mainly focused on aboveground biomass (AGB). AGB plays an important role in carbon fixation, and is a vital surrogate of forest ecosystem functioning [9]. As a surrogate of forest structure, we used the growing space occupied (GSO) as an indicator of ecological resilience. The GSO is the growing space occupied by species of a specific site, and is commonly used as the crown closure measurement. Growing space primarily reflects forest structure, and can distinguish forest successional stages over large landscapes [32]. Seidl et al. [9] selected the rumple index (RI) of canopy complexity as a surrogate of vegetation structure. The rumple index is the ratio of the canopy surface area to the projected surface ground area [33]. The RI was proposed as a powerful composite index to describe vegetation structure and distinguish different stages of forest development over large areas [34], which was similar to the contents of GSO. Thus, we used GSO as the indicator of ecological resilience in this study. The measure of GSO incorporated in LANDIS PRO is calculated by the following equations:

$$GSO_i = \sum_{j=1}^{longevity/timestep} \left(\left(\frac{DBH_j}{10\ inch} \right)^{1.605} \times NT_j \times \frac{1\ hectare}{MaxSDI_i \times site\_area} \right)$$

$$GSO = \sum_{i=1}^{number\ of\ species} GSO_i$$

where DBH_j and NT_j are the mean diameter and number of trees of j th diameter class for species i in inches and stems, respectively; MaxSDI is the maximum stand density index (derived from species vital attributes). Stand density index (SDI) is a basic concept in forestry. It was first developed by Reineke in 1933 and has been widely used to characterize stand density (tree crowding). Growing Space Occupied (GSO) represents an extension of SDI to meet the needs of landscape modeling, and is also a key metric within LANDIS PRO [32].

As a surrogate of vegetation component, we selected the percentage of older age cohort individual species (ACS). Forests which contain tree species with diverse age and size structures have been observed to be relatively more resilient to climate change and fire disturbance than forests with younger, less diverse age structures [31]. ACS is defined by the ratio of old age cohorts to the total trees cohorts (trees age > 60 year for broadleaf, and >100 year for conifers). For the surrogate of composition, we selected the proportion of mid, and late-seral species (i.e., larch, Mongolian Scots pine, and Korean spruce) >5 cm in DBH (diameter at breast height) (LSS).

Table 1. Indicators of forest ecological resilience.

Property	Indicators	Description	Calculation Methods
Functioning	AGB	Aboveground biomass	Derived from LANDIS PRO model outputs
Structure	GSO	Growing space occupied	The percentage of growing space on the site occupied by all species
Composition	ACS	Age cohort structure	$\frac{\text{broadleaf trees age} > 60\ \text{yr} + \text{conifer trees age} > 100\ \text{yr}}{\text{the nmmber of all trees in each cell}}$
	LSS	Proportion of mid- and late-seral species	$\frac{\text{all conifer trees in each cell}}{\text{the number of all trees in each cell}}$

In order to facilitate the comparison and calculation among different indicators, we first used the min-max normalization method to normalize all four indicators (AGB, GSO, ACS, and LSS) by each time step at both land type and landscape scales. The data normalization process is calculated using the following Formula (1):

$$\overline{X_i} = \frac{|X_i - X_{min}|}{X_{max} - X_{min}} \quad (1)$$

where $\overline{X_i}$ is the normalized data, which ranging from 0 to 1. X_i is the value of the indicator at year *i*. X_{min} and X_{max} represent the minimum and maximum values, respectively.

We then calculated the ecological resilience at all simulated time steps by using all four indicators (AGB, GSO, ACS, and LSS). The calculated ecological resilience is a specific number ranging from 0 to 1 that quantifies the capacity of different forest stands to recover from extra disturbance and maintain a stable status. Forest stands with high resilience values have a higher ability of recovery (more resilient than other stands). The formula for this is (2):

$$R_i = W_1 \times \overline{AGB_i} + W_2 \times \overline{GSO_i} + W_3 \times \overline{ACS_i} + W_4 \times \overline{LSS_i} \quad (2)$$

where R_i is the ecological resilience value at year *i*, higher R_i means higher ecosystem recovery ability (ecological resilience). $\overline{AGB_i}$, $\overline{GSO_i}$, $\overline{ACS_i}$, and $\overline{LSS_i}$ represent the normalized AGB, GSO, ACS, and LSS values at year i, respectively. W_j (j = 1, 2, 3, 4) are the weight coefficients, which are calculated by using the coefficient of variation method in the following Formula (3):

$$W_j = \frac{\sigma_j}{\overline{\overline{X}}_j \times \sum_{j=1}^{n} \frac{\sigma_j}{\overline{\overline{X}}_j}} \quad (3)$$

where W_j is the weight coefficient of indicator j; σ_j is the standard deviation of indicator j; $\overline{\overline{X}}_j$ is the mean value of the indicator j.

2.3. Simulation Experiments Design and Data Analysis

We designed a factorial experiment to assess the effects of climate change, climate-induced fire regimes, and forest management schemes on boreal forest ecological resilience. In this factorial experiment, we set three independent variables: Climate change (current climate and climate change), different fire regimes (current fire and climate-induced fire), and forest management schemes (no treatment and different harvesting and planting strategies).

The current meteorological data were derived from the meteorological center, and monthly temperature and precipitation data were included from year 1961 to 2000. We used the data derived from five related weather stations to build regression models among spatial positions, elevations, and temperature as well as precipitation in the studied area. We then calculated the mean annual temperature and precipitation of different land types by using this regression model. We used two different levels of carbon emissions scenarios (CGCM3 B1 and UKMO-HadCM3 A2) to represent future climate change in our study. The B1 scenario represents low CO_2 emissions, while A2 represents high CO_2 emissions [3,35]. Based on the projected data of Hadley GCM, the mean annual temperatures and precipitations would increase linearly in year 2000-2100, and after that it would enter into a stable state [36]. The historical fire regimes for our simulations were characterized by the Chinese Federal Forest Service database from 1965 to 2005. According to previous study, fire occurrences in our study region under the B1 and A2 scenarios (projected by the Hadley GCM) would increase by 30% and 200% compared to historical fire regimes, respectively [3].

We used recent harvest trends in our study area to construct the current harvest regime. To examine the effects of the current harvest regime and future possible forest management schemes on forest ecological resilience, we designed eight harvesting and planting scenarios (Table 2). These scenarios include a combination of designated harvest intensity and increasing percentages of individual trees planted to the current intensity (P0) to 10% (P10), 20% (P20), 30% (P30), 40% (P40), and 50% (P50) of the mean stand density.

Table 2. The scenarios for different harvesting and planting strategies (HP) simulated by LANDIS PRO.

Scenarios	Harvesting and Planting Were Permitted	
	Harvesting Intensity for Species (H)	**Planting Intensity of Conifer Trees (P)**
H1P0	Cut conifer trees only	0 for each planted species
H2P0	Cut broadleaf trees only	0 for each planted species
H3P0	Cut broadleaf and conifer trees	0 for each planted species
H4P10	Cut broadleaf trees	10% for each planted conifer species
H4P20	Cut broadleaf trees	20% for each planted conifer species
H4P30	Cut broadleaf trees	30% for each planted conifer species
H4P40	Cut broadleaf trees	40% for each planted conifer species
H4P50	Cut broadleaf trees	50% for each planted conifer species

Specifically, we designed five simulated scenarios: (1) Current fire only scenario (CF1: the reference scenario, fire and succession were simulated with current fire occurrence); (2) Increased fire only scenario (CF2: compared to current fire regime, fire occurrence increased by 30%; CF3:

fire occurrence increased by 200%); (3) climate change only scenario (B1F1 and A2F1: climate change and current fire regimes were simulated); (4) climate-induced fire scenario (B1F2: B1 climate and fire occurrence increased by 30%; A2F3: A2 climate and fire occurrence increased by 200%); and (5) climate-fire- forest management schemes (B1F2HP: B1 climate, fire occurrence increased by 30% and harvesting and planting; A2F3HP: A2 climate, fire occurrence increased by 200% and harvesting and planting). We used a FLM to simulate 5 tree species (Table S1) at 5-year time step from year 2000 to 2300 with five replicates to reduce the stochasticity.

To examine the effects of extra factors on boreal forests, we compared the response variable, ecological resilience, under the reference scenario to the scenarios climate change only, increased fire only, and climate induced-fire, and climate-fire-forest management schemes for short- (0–50 year), medium- (50–150 year), and long-term (150–300 year) simulation periods using the mean comparison method. An analysis of variance (ANOVA) was used to test the differences between the reference scenario and all other scenarios. We used the Tukey's Honestly Significant Difference (HSD) method for post-hoc analyses at all simulated periods. To evaluate the increased fire effects on boreal ecological resilience, we tested the response variable among different fire regimes at all three simulated periods. All statistical analyses were conducted using SPSS 23.0 software.

2.4. Simulating Ecological Resilience from Climate Change and Disturbance

We employed a forest landscape model to simulate forest succession under different climate, fire regimes and forest management schemes, and to evaluate ecological resilience of boreal forests (Figure 2). LANDIS PRO is a spatially explicit landscape model, and can be used to simulate forest dynamic over large spatial and temporal scales with user defined resolutions (10–500 m) [34]. LANDIS PRO records density and size information for each age cohort by species within raster cells enabling the model to directly output spatially explicit stand information (e.g., density, basal area, and aboveground biomass). This data structure enables forest inventory data to be directly used for model initialization and parameterization. LANDIS PRO can simulate tree growth, species establishment, mortality, and species resources competition within each raster cell, and also simulate seed dispersal, forest management, and natural disturbance across the whole landscape. LANDIS PRO stratifies the entire landscape into relatively homogenous land type units based on climate, soil, terrain, and other environmental factors. Species establishment probability (SEP) is a key input parameter of LANDIS PRO. SEPs are obtained based on responses of each species to specific microenvironment factors such as soil moisture, soil N, soil C, and local climate. LANDIS PRO uses SEPs as inputs to indirectly capture the spatial variability of climate. Species with high SEPs have a higher probability of establishment. The SEPs of specific species are derived from previous LANDIS modeling studies or a gap model (e.g., LINKAGES).

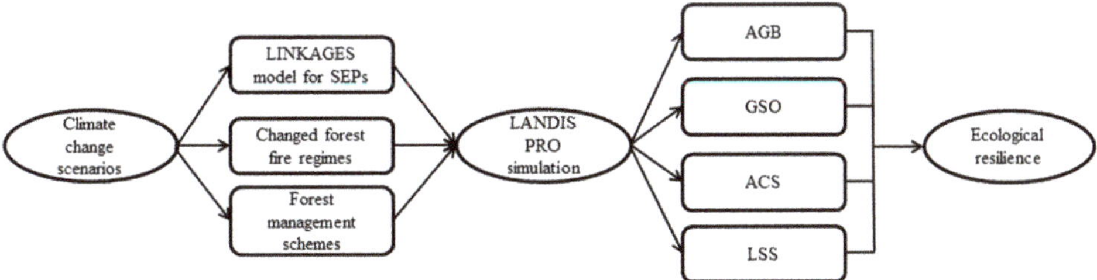

Figure 2. The framework for model-coupling and sub-methods used to evaluate ecological resilience. SEPs: species establishment probability, AGB: aboveground biomass, GSO: the growing space occupied, ACS: age cohort structure, and LSS: proportion of mid and late-seral species.

In the fire module, fire disturbances are simulated based on specific input parameters (e.g., mean fire return interval, mean fire sizes) [37]. Fire disturbance is simulated within spatially defined fire regime units which have parameterized ignition rates, fire size distributions, and mean return intervals. The fire module includes three major components which are fire occurrence, fire spread, and fire effect simulation. Forest fires effects are characterized by bottom-up disturbance, and younger trees are more susceptible to fire than older ones in model simulation.

In the harvest module, forest harvesting is simulated using a management area map and forest stand map. The management area map and the stand map provide boundaries for specific harvest events to occur. Clear-cutting, thinning (from above or below), and group selection harvesting can be specified to execute harvest events in the harvest module [38]. By using those parameters, many common forest management schemes can be simulated in this module.

2.4.1. LANDIS Model Initialization and Parameterization

Parameterization LANDIS PRO included two aspects: Non-spatial parameters (species' vital attributes, SEPs, site-level fire disturbance, harvest scenarios, and planting parameters), and spatial parameters (species composition map, land type map, management area map, stand map, and fire regime unit map). Five of the most common tree species were simulated in LANDIS, which account for more than 90% of the total forested land [29]. Species' vital attributes were derived from previous studies in the same or similar regions, field investigation, and consultation with local experts [39,40]. We used land use data, Landsat imagery, slope, and aspect maps to classify the study area landscape into six land types. We then used LINKAGES to simulate the response of forest species under current and climate change scenarios within each land type, and used the simulated individual species biomass to estimate SEPs for each simulated species. We modeled the SEPs of five tree species under different climate conditions by each land type (Table 3). The initial SEPs were estimated by current climate (1961–2000), and the SEPs for future scenarios were projected by climate change data (2010–2099). The SEPs were assumed to change linearly in 2000–2100, and held constant after 2100. Specifically, we used LINKAGES to simulate the individual biomass of different tree species to both current and climate warming scenarios within each land type. The individual species biomass was used to estimate the SEPs for specific species. The SEPs are calculated by the following equation:

$$SEP_{ij} = \frac{b_{ij},\ b'_{ij}}{max\left\{\sqrt{\sum_{j=1}^{n} b_{ij}},\ \sqrt{\sum_{j=1}^{n} b'_{ij}}\right\}}$$

where b_{ij} and b'_{ij} are the biomass of species i on land type j under current and warmer climate, respectively, SEP_{ij} is the species establishment probability of species i on land type j under current and warming climate [36,41].

The forest composition map was obtained based on forest stand maps, forest inventory data, and field data, which included species spatial location, number of trees and age cohort information. The forest stand map was acquired in the 2000's was a GIS based file, that provided stand site boundaries, species composition, structure, and the average age of the specific polygons. We derived sample plots investigated during the 2000's, which provided number of trees and age cohorts by species from the China National Forest Inventory Second and Third Tier data. We integrated the forest stand map (vector format) and forest inventory data (stand information) to derive the initial forest composition map. To reduce computational resources, all those input maps needed by LANDIS model were converted to a resolution of 90 m × 90 m cell size (2217 columns × 2609 rows) by using ESRI ArcGIS software.

Fire regime parameters and a fire regime unit map were required for the fire module. The fire regime unit map was used to identify areas with heterogeneous fire properties across the landscape, and fire characteristics in this region were mostly related to soil moisture, terrain, climate, and vegetation traits, which were closely related to the classification of land types, and thus we used the land type map as the fire regime unit map in our study area. The current fire occurrence for our simulations was parameterized based on data from the historical fire database recorded from 1965 to

2005. Based on the database, we calculated the current fire regime parameters (e.g., return interval, ignition rate, and mean fire size) of each fire regime unit. The future fire regimes were characterized by changing fire occurrences under different climate scenarios based on previous work [3]. The boreal forests in our study area have been exploited since the 1950's, and timber harvesting has extensively altered forest composition, structure, and age cohort. Consequently, to maintain forest ecosystem function and sustainability, timber harvesting has been restricted by a natural forest conservation project since 1999. Mongolian Scots pine and Korean spruce were extensively cut because of their high economic value and stands typically reestablished with larch. At present the local forestry bureaus have attempted to actively protect the remaining stock of these two species. In accordance with current harvest policy, the harvested species were larch, birch, and aspen, whereas pine and spruce were not harvested. The predominant harvest type in our study area was thinning from below, and all harvest scenarios were processed by removing the smallest trees first. We simulated the current harvest activities by using a basal area controlled harvest method (tree species were removed from a stand until a specific target basal area value was reached) followed by planting in permitted areas.

Table 3. SEPs for each available land type under current climate and climate change scenarios.

Land Type	Climate Scenario [1]	Species Establishment Probabilities (SEPs)				
		Larch	Pine	Spruce	White Birch	Aspen
Terrace	C	0.200	0.050	0.050	0.030	0.070
	B1	0.060	0.200	0.180	0.076	0.166
	A2	0.000	0.000	0.000	0.418	0.186
Southern slope	C	0.350	0.350	0.005	0.350	0.030
	B1	0.376	0.327	0.174	0.284	0.106
	A2	0.141	0.320	0.111	0.669	0.271
Northern slope	C	0.400	0.010	0.030	0.150	0.005
	B1	0.522	0.406	0.388	0.190	0.042
	A2	0.270	0.151	0.245	0.238	0.213
Ridge top	C	0.200	0.010	0.000	0.070	0.020
	B1	0.346	0.100	0.020	0.076	0.010
	A2	0.413	0.325	0.180	0.222	0.147

[1] C: current climate condition; B1 and A2: climate change scenarios.

2.4.2. Model Calibration and Verification

Simulated results (e.g., species composition, tree density, basal area, and aboveground biomass by species for each cell and time step) from LANDIS PRO can be directly compared with forest inventory data as a method of calibration and validation [32]. In order to parameterize the initial forest landscape accurately, we used 70% of the inventory plots (investigated in 2000s, consists of the number of all trees and age cohorts) and the stand map (a GIS file) to initialize the forest composition map at year 2000, and then simulated the model for ten years. We iteratively adjusted species' growth curve (an essential input parameter used to control tree growth and calculate species biomass) to make the initialized forest stand information match the remaining 30% of the forest inventory data at year 2000. We then calibrated the number of potential established seeds (a parameter related to tree density and basal area) until the simulated results for year 2010 was similar to field data for the year 2010. This calibration ensured that species' growth curves and the number of potential established seeds was suitable for our study area [39]. To evaluate the simulated landscape at year 2010, we used a scatter plot of the observed density and basal area vs. the simulated density and basal area. We first selected 322 raster cells from the simulated landscape at year 2010, and then the density and basal area were extracted from selected cells to compare with forest inventory data. Likewise, the forest inventory data (322 plots, investigated in 2010s) were also converted to the total density and basal area.

To verify simulated fire on the forested landscape, we compared the model results with field data at different simulated periods. We ran the current fire only scenario (CF1) for 300 years, and randomly selected 40 fires with low intensities (more than 90% fires occurred at this level in our study area) from different years and locations from the LANDIS PRO output. We then inventoried 40 field sites (8 field sites per each age group) that were actually burned 5, 10, 15, 20, and 25 before the year they were sampled. We set five plots with 20 m × 20 m in each field site. We then recorded all the individual trees with basal diameter above 1 cm level in each plot. Tree number and DBH by species were measured at each plot, and these plot data were converted to density (trees/ha) and basal area (m^2/ha). We statistically compared tree density and basal area of the 40 simulated fires with 40 corresponding fires sampled in the field, respectively.

To ensure the simulated results more authentic, we also compared the simulated aboveground biomass to previous studies, which conducted plot surveys in similar region at landscape scales. We used the currently available data for model evaluation. While predicted results under climate change scenarios over next 280 years cannot be verified by filed inventory data, simulated successional and stand dynamic trends have been confirmed by other studies conducted at similar regions [36,42].

3. Results

3.1. Model Calibration and Validation

Our simulated results indicated that the initialized forest composition constructed from the observed data from year 2000 adequately represented the forest landscape (stand density: $R^2 = 0.821$, Pearson correlation test: $p < 0.01$; basal area: $R^2 = 0.804$, $p < 0.01$) (Figure 3a,b). The simulated stand density and basal area were close to the observed forest inventory data at year 2010 (stand density: $R^2 = 0.803$, $p < 0.01$; basal area: $R^2 = 0.832$, $p < 0.01$) (Figure 3c,d). Thus, we accepted the calibrated results for further calculation.

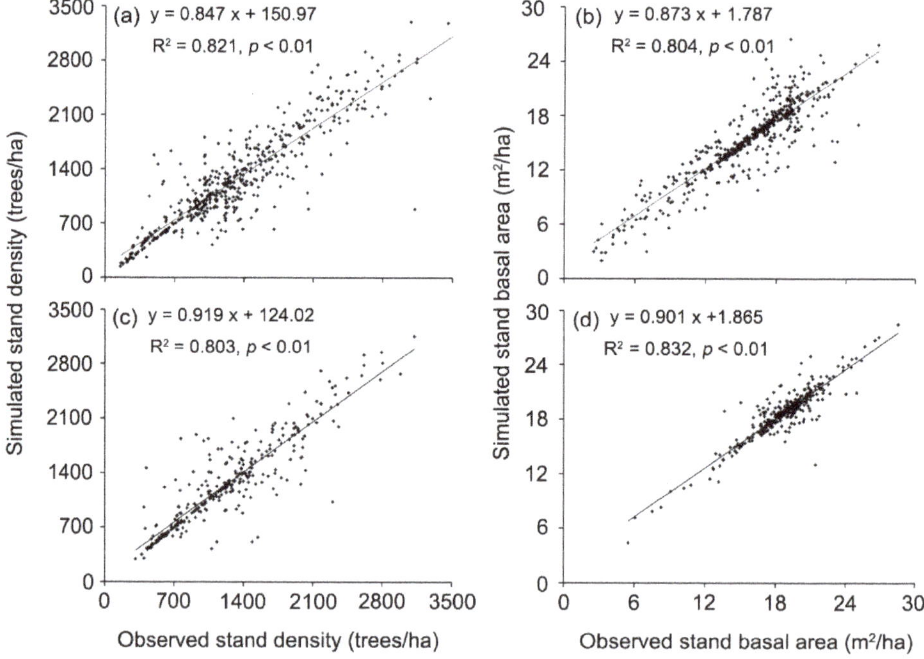

Figure 3. Scatter plot showing the relation between simulated and observed data for stand density (**a**,**c**); and basal area (**b**,**d**) at year 2000 and 2010, respectively (Pearson correlation test: $p < 0.01$).

The results showed that the post-fire stand density increased during the first 10 years (up to 17,295 trees/ha) and then decreased to 4725 trees/ha after 25 years (Figure 4a). The increasing trend was largely attributed to forest fires removing many trees causing the release of growing space for pioneer species to establish. After year 10, these post-fire stands reached the self-thinning stage, and began to reduce individual trees in the following years. The post-fire basal area showed an increasing trend throughout the observed 25 years (Figure 4b). The simulated trends in both stand density, basal area and aboveground biomasses closely followed trends in the field sample data (Figures 4 and 5).

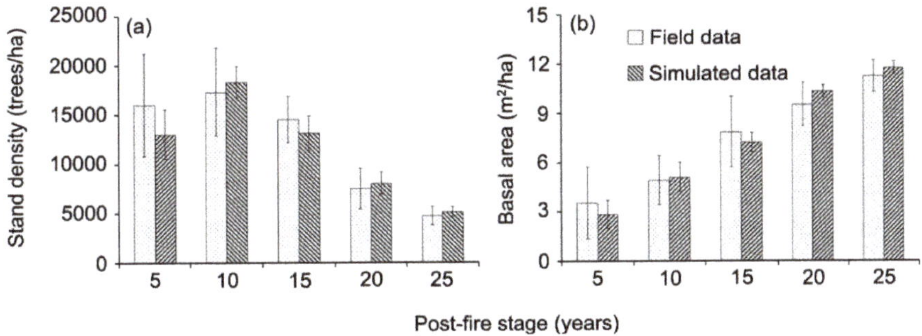

Figure 4. Comparison between observed and simulated stand density (**a**) and basal area (**b**) in burned areas in relation to post-fire year.

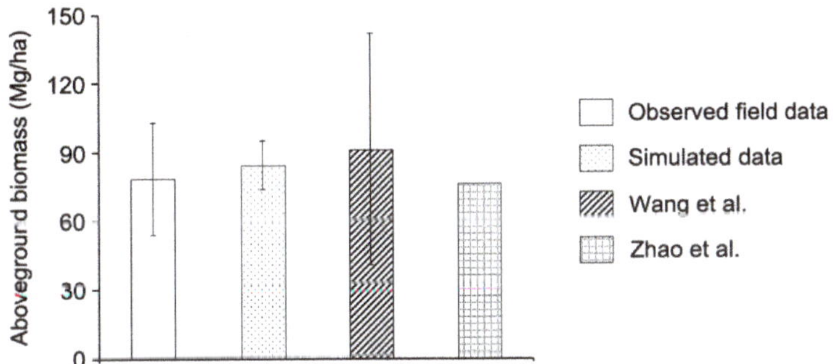

Figure 5. Comparison between simulated data and observed field aboveground biomass, published data from plot surveys in the study area.

3.2. Ecological Resiliencies Response to Climate Change and Fire Regimes at Landscape Scale

Our results showed that under current climate conditions, ecological resilience was affected by forest fire regimes (Figure 6a). The forest ecological resilience was greatest under the CF2 scenario followed by the CF1 and CF3 scenarios. Under the current fire regimes (CF1), the ecological resilience increased rapidly from the simulated years 50 to 80, then decreased until year 160, and then increased slightly to year 300. The ecological resilience decreased significantly under the CF3 scenario in the first 160 years compared to the CF1 scenario. However, the ecological resilience under CF1 scenario coincided with the CF3 scenario from year 170 to 270.

The trajectories of forest ecological resilience varied among climate change and fire regime scenarios (Figure 6b). Under B1F1, B1F2, and A2F1 scenarios, the ecological resilience dynamics had a similar trend for the whole simulation period. The curves of these three scenarios fluctuated in the first 50 years and peaked at year 80, then decreased until year 160, and then increased to year 300 gradually. The calculated forest ecological resilience was highest under the B1F2 scenario across all simulated periods, followed by B1F1, A2F1, and A2F3 scenarios. Moreover, the ecological resilience under the A2F3 scenario was visibly lowest among these scenarios until year 210.

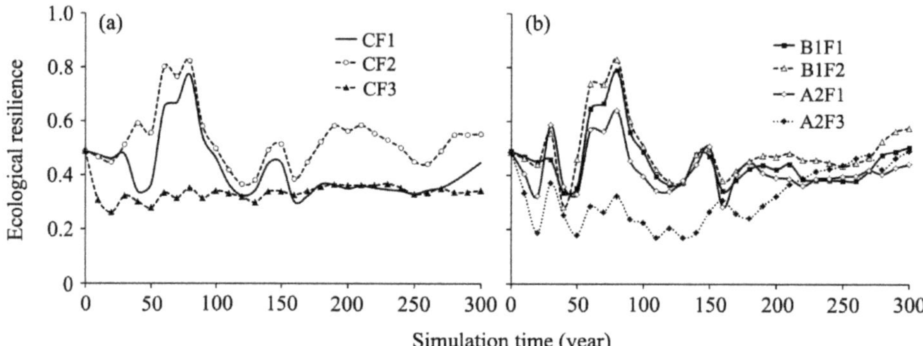

Figure 6. Changes in ecological resilience at the landscape level under different simulated scenarios. (**a**) current climate condition with different fire occurrence (CF1, CF2, and CF3); and (**b**) climate change scenarios with different fire occurrences (B1F1, B1F2, A2F1, and A2F3).

3.3. Effects of Fire Regimes, Climate Change on Ecological Resilience

The effects of forest fire on ecological resilience varied among the three fire occurrence scenarios across the three simulated periods (Figure 7). Compared to the CF1 scenario, no significant difference was found between the short and medium term interval ($p > 0.05$) under CF2 scenario, while the ecological resilience under the CF2 scenario differed significantly from CF1 scenario during long-term interval ($p < 0.05$). However, the ANOVA tests demonstrated that the simulated ecological resilience under CF3 scenario for both short and medium term interval differed significantly from CF1 scenario ($p < 0.05$), and no significant difference existed in the long term interval (150–300 year).

The ecological resilience was substantially higher under the CF2 scenario than the CF1 scenario (Figure 7). Our results showed that the increase in ecological resilience under CF2 scenario was 17.5%, 12.4%, and 43.2% greater than that in CF1 scenario across the entire simulated periods, respectively. Under CF3 scenario, the largest reduction in ecological resilience occurred in the short and medium term interval, and was 24.6% and 34.3% lower than that in the CF1 scenario. However, the average value of ecological resilience under CF3 scenario was similar in the long-term period.

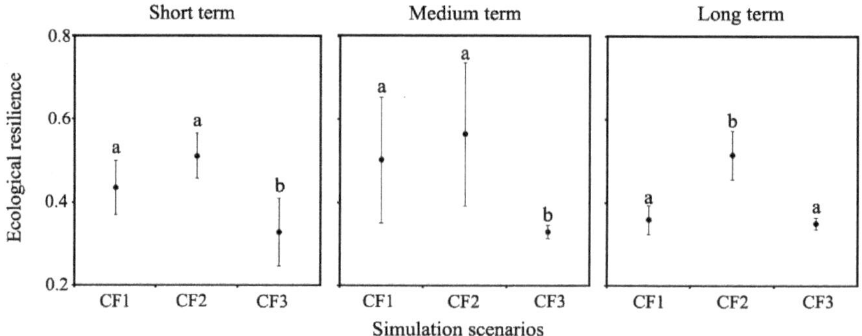

Figure 7. Multiple comparison of three fire scenarios effects on ecological resilience at different simulated periods. Small letters indicated significant differences among scenarios at 0.05 level. CF1, CF2, and CF3: different fire occurrence scenarios.

The ecological resilience responses to three climatic scenarios differed among different land types (Figure 8). Results showed that the ecological resilience increased more sharply in the first 80 years (up to 0.919), then decreased until year 170, and then increased slightly to year 300, and the ecological resilience responded negatively to both B1, A2 climate scenarios after year 80 on the terrace land type (Figure 8a). On south-facing slopes, the ecological resilience decreased in the first 50 years, then increased slightly until year 80, and remained almost stable afterward under the CF1 and B1F1 scenarios. Ecological resilience under A2F1 scenario decreased by 61.5% in the first 130 years, and then generally increased to year 300 (Figure 8b). The curves of ecological resilience in north-facing land type responded to climate change similar to that on south-facing land types (Figure 8c). On the ridge top land type, the curves of ecological resilience under three climatic scenarios decreased sharply in the first 30 years, and then increased slightly afterward (Figure 8d). There was a slight increase of ecological resilience in B1F1 and A2F1 scenarios compared to the CF1 scenario after year 120 (Figure 8e).

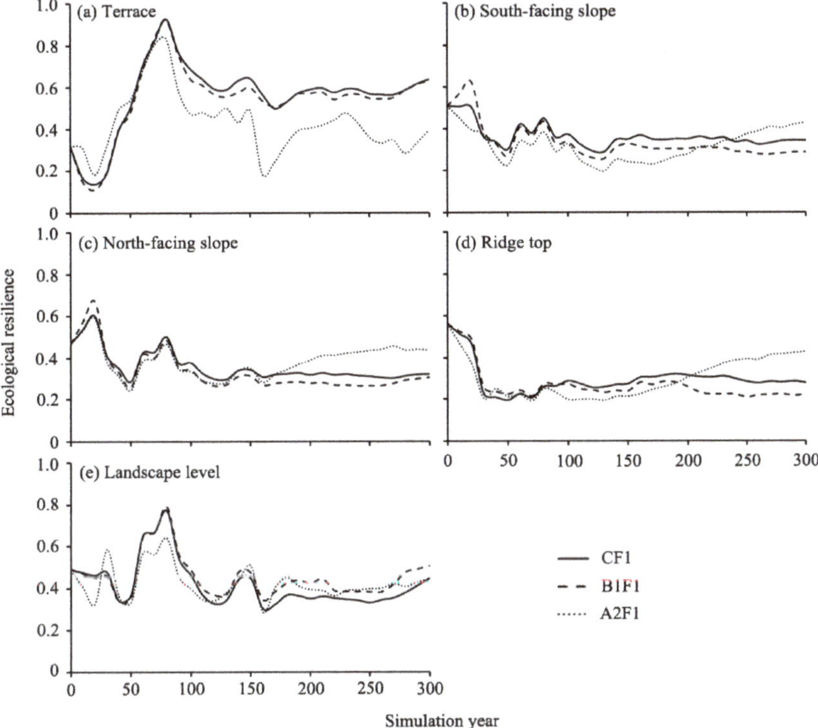

Figure 8. Ecological resilience dynamics for different land types under three climatic simulated scenarios. CF1: current fire only scenario, and B1F1, A2F1: climate change only scenarios.

3.4. The Interactive Effects of Fire Disturbance and Climate Change on Ecological Resilience

There was no significant difference of ecological resilience on the terrace land type between CF1 and B1F2 scenarios during the simulated periods ($p > 0.05$, Figure 9); and ecological resilience on the terrace land type decreased by 1.7%, 5.9%, and 1.7% at the three simulated periods compared to the CF1 scenario. For the South-facing and North-facing land types, the ecological resilience did not differ significantly between CF1 and B1F2 scenarios for the short and medium-term interval ($p > 0.05$, Figure 9). However, ecological resilience differed significantly between the B1F2 scenario and the CF1 scenario for the long-term period ($p < 0.05$, Figure 9), where ecological resilience was 11.3% lower than that in CF1 scenario. On the ridge top land type, no significant difference was found between CF1 and B1F2 scenarios for the whole simulation periods, where ecological resilience was 14.1% higher, and 3.7% lower, respectively, than that in the CF1 scenario. The results showed that ecological resilience under the B1F2 scenario was significantly higher than that under CF1 scenario at landscape scale only for the long-term period ($p < 0.05$, Figure 9). The increase in ecological resilience at the landscape scale was 3.2%, 12.1%, and 29.6% greater than that in the CF1 scenario during the three simulated periods, respectively.

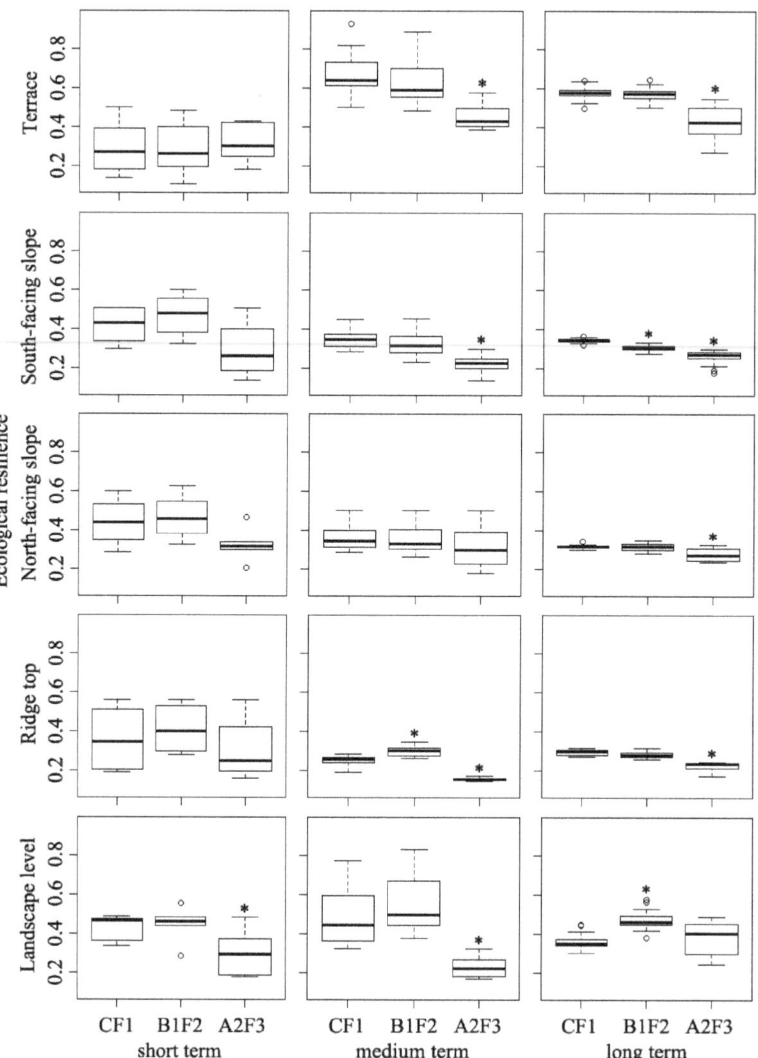

Figure 9. Ecological resiliencies of different land types for the three simulated scenarios: current fire only scenario (CF1); climate change induced-fire scenarios (B1F2, A2F3). Short term: 0–50 years, medium term: 50–150 years, and long term: 150–300 years; * indicates that significant differences are detected between the CF1 scenario and a given scenario ($p < 0.05$).

Ecological resilience under the A2F3 scenario differed significantly on terrace, South-facing slope, and ridge top land types from CF1 scenario at medium and long-term periods ($p < 0.05$, Figure 9), and no significant differences were detected at short-term interval on those three land types. On North-facing slopes, the ecological resilience under A2F3 scenario differed significantly from CF1 scenario during the long-term period ($p < 0.05$, Figure 9), and the decrease in ecological resilience was 26.7%, 12.4%, and 12.3% lower than that in the CF1 scenario across the three simulated periods, respectively. Collectively, our results indicated that the B1F2 scenario did not affect ecological resilience across the short- to medium-term range, but with continuous climate and fire influence, it will significantly affect the ecological resilience at landscape level across the long-term range. Meanwhile, under the A2F3 scenario, forest ecological resilience could be recovered almost to its original state by 150 years simulation time.

3.5. The Effects of Forest Management Schemes on Ecological Resilience

Our results showed that forest management schemes played a role in altering ecological resilience under the climate change scenarios in contrast to the no management treatments (Figure 10). For the B1F2 scenario, ecological resilience decreased obviously under the eight harvesting and planting scenarios (Figure 10a). Under all simulated scenarios, ecological resilience initially fluctuated, then peaked at year 80, and then decreased until year 130, and increased gradually afterward. The curves of ecological resilience for the eight harvesting and planting scenarios were relatively lower than that under B1F2 scenario during most of the simulation periods. Generally, our results indicated that all eight harvesting and planting strategies did not affect ecological resilience at landscape scale under the B1F2 scenario during all simulation periods.

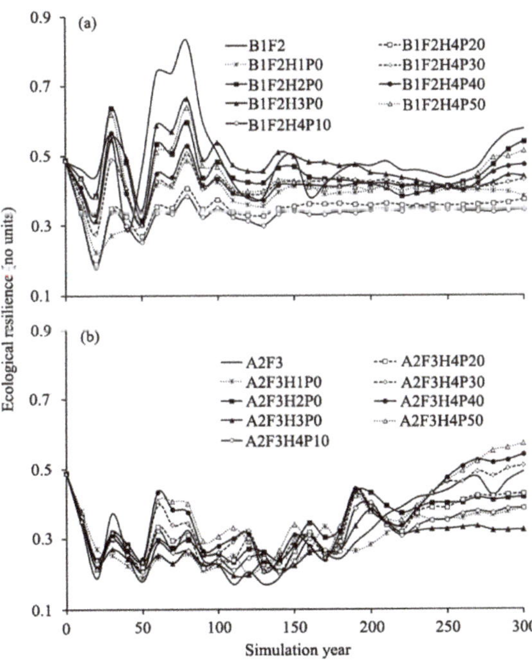

Figure 10. The ecological resilience simulated under different forest management schemes at landscape scales under two climate change scenarios in comparison with no management treatment (B1F2 and A2F3): (**a**) The ecological resilience affected by harvesting and planting in B1 climatic scenario; and (**b**) the ecological resilience affected by harvesting and planting in A2 climatic scenario.

For the A2F3 scenario, ecological resilience also varied among these eight harvesting and planting strategies across all simulation periods (Figure 10b). Ecological resilience decreased sharply under all the harvesting and planting scenarios in the first 20 years, and then fluctuated until year 220 and increased gradually afterward. The ecological resiliencies of eight harvesting and planting scenarios were higher than that under A2F3 scenario during the first 190 years simulation. Under A2F3H4P30, A2F3H4P40, and A2F3H4P50 scenarios, the ecological resiliencies were higher than that under the A2F3 scenario after year 240 (Figure 10b). Our results indicated that certain harvesting and planting strategies needed to be implemented under the A2F3 scenario, and only three of the eight harvesting and planting strategies affected the ecological resilience positively at the landscape level during the simulation periods except for simulated year 210 to 240.

4. Discussion

Boreal forest ecosystems have the ability to recover from disturbances without undergoing fundamental change (a quality referred to as resilience) that it is dependent on functions and structures at multiple scales of the forest ecosystem [15,43]. Understanding and quantifying responses of ecological resilience to extra disturbances is a challenge [44] because ecological resilience dynamics are an inherent ecosystem property that is related to multiple scales and comprehensive spatiotemporal data does not exist, and further cannot be feasibly measured directly by field observations [16,45]. Ecological resilience can be estimated by means of ecological resilience indices such as biodiversity, habitat conditions, and productivity etc., and many studies had been focused on this issue [10,46,47]. For example, Scheffer et al. [20] suggested that maintaining the ecological resilience of forest ecosystems was likely be the most feasible and effective way to manage forest ecosystems under possible future changing environments. Chapin et al. [15] assessed the resilience of boreal forest ecosystems to rapid climate change. In contrast to those studies, the spotlight of our study lies in the quantitative parts and resilience prediction. To our knowledge, few previous studies have used resilience indicators to calculate the ecological resilience at a landscape scale, and evaluate the effects of climate change and fire regimes on the ecological resilience at both land type and landscape scales. In our research, we used a FLM to evaluate the ecological resilience of forests to climate change and altered fire regimes. This modeling approach can be used to further explore these broad-scale issues, and to evaluate the interactions of forest succession and disturbance dynamics. Furthermore, it provides long term insights in exploring forest ecological resilience. The LANDIS PRO model can explicitly track aboveground biomass, species composition, tree number, and age cohorts, which can be verified by directly comparing to the field data. The validation process conducted in our study added to the robustness of our modeling approach and confidence of predictions (Figures 3–5). To verify the aboveground biomass indicator, we compared our results with field data and published data in the same or similar regions. The results showed that the predicted aboveground biomass (84.9 $\pm$ 10.6 Mg/hm^2) were within the observed ranges of the field sample data we collected (78.5 $\pm$ 24.4 Mg/hm^2) and the published data reported by Wang et al. [48] (91.4 $\pm$ 50.4 Mg/hm^2) and Zhao et al. [49] (76.5 Mg/hm^2), respectively (Figure 5). Utilizing the full range of outputs from the LANDIS PRO model make it possible to further explore the dynamics of ecological resilience, and to assess the effects of climate change, climate-induced fire regimes, and forest management schemes.

Ecological resilience was enhanced under the low fire occurrence scenario in comparison to the current fire occurrence condition, whereas it decreased under the high fire occurrence scenario (Figure 7). This suggested that low fire occurrence had positive effects on boreal forest's resilience, while high fire occurrence should be avoided in future forest management. However, previous studies showed that low fire occurrences had negative effects on boreal forest ecosystem resilience [36,50]. This difference may be related to response variable selection or analysis of fewer indicators of ecological resilience. Meanwhile, the curve of ecological resilience under the CF2 scenario increased in the first 80 years and decreased in the next 50 years. This was likely because low-intensity forest fires removed mostly small trees and released growing space for white birch and aspen to recruit. After the process of post-fire tree recruitment progressed over the first 70 to 80 years, self-thinning began to cause mortalities of pioneer species over the next 50 years. Meanwhile, many of these pioneer trees had reached their longevity and began to die in the self-thinning process [51].

Our results revealed that ecological resilience under climate change scenarios changed differently among simulated land types. Under climate change only scenarios, the curves of ecological resilience were slightly lower than that under current fire only scenario among different land types during different simulated periods (Figure 8). This may have been related to the variation in species establishment probabilities among climate scenarios at different land types, and the initial forest composition and tree distribution [22,52]. Within the terrace land type, the ecological resilience was significantly lower under A2F1 scenario than that under CF1 and B1F1 scenarios after year 80 (Figure 8a). This was because most of the terrace land type area was covered by coniferous species with trees in middle-age cohorts, and the present-day dominant larch trees could not establish under A2 climate scenario [29]. There was a time lag for the effects of climate change and fire disturbance on ecological resilience among different land types (Figure 9). This was consistent with previous studies [36,53]. Furthermore, our simulated results indicated that time lags were varied among different land types under climate change scenarios. For instance, the response time of ecological resilience to A2F3 scenario was 150 years in North-facing slope region, which was almost 100 years longer than that in the South-facing slope land type, and this discrepancy may related to the distribution patterns of solar energy and available water resources in future climate change scenarios between these two land types, and the current coniferous and broadleaf species distribution [54].

Forest management schemes played an important role in influencing ecological resilience under climate change scenarios at the landscape level (Figure 7). Our results showed that many of the harvesting and planting strategies had negative effects on the ecological resilience compared to the B1F2 scenario. There were two reasons for this: (1) Fire regimes under B1F2 scenario removed many small trees, and released growing space for species to occupy. Meanwhile, the thinning methods of harvesting also removed most small broadleaf trees; (2) Due to shortages of growing space, planting efficiency of coniferous trees was relatively low, and the planted trees did not offset total removals due to fire and harvest events [29]. In this light, we concluded that no additional forest management treatments were suitable under the future B1F2 scenario. Our results moreover showed that most of the harvesting and planting strategies had positive effects on ecological resilience, and the curves of ecological resilience under A2F3H4P30, A2F3H4P40, and A2F3H4P50 scenarios were obviously higher than that under the A2F3 scenario in most of the simulated periods. This may have been related to the influences of changed environmental conditions under A2 climate scenario, increased fire occurrence, and the different biophysical limits of coniferous and broadleaf trees when facing future changing climates [55]. This suggested that the three of eight strategies were suitable under the future A2F3 scenario.

5. Conclusions

In this study, we predicted the dynamics of forest ecological resilience indicators (AGB, GSO, ACS, and LSS) at both landscape and land type scales in boreal forests by employing a forest landscape model, and then quantified the ecological resilience by incorporating those representative indicators. This modeling approach also provided insight into ecological resilience trends under changing climate conditions, fire regimes and possible future forest management schemes. In conclusion, we found that: (1) The LANDIS PRO model can be implemented in evaluating ecological resilience of boreal forests at multi scales in Northeastern China; (2) the ecological resiliencies of forests in the Great Xing'an mountains were likely to be significantly altered by different climate conditions, fire regimes, and their interactive effects during most of the simulated periods; (3) the direct effects of climate variations on forest ecological resilience in the study area are not likely to be as important as the possible changed fire regimes at the landscape scale, and future climate warming (high CO_2 emission) with high fire occurrence regime would significantly reduce the ecological resilience of forest ecosystem;

(4) the proposed forest management schemes do not mitigate the effects of climate variation and climate-induced fire regime effects under the low climate-induced fire regime scenario, and medium and high intensities of forest management schemes (30%, 40%, and 50% intensities) are proposed under the high climate-induced fire regime scenario. These results provided useful information for future boreal forest managements.

Supplementary Materials: The following are available online at http://www.mdpi.com/2071-1050/10/10/3531/s1, Table S1: Main species attributes of our study landscape.

Author Contributions: X.L. and H.S.H. designed the simulated scenarios, analyzed the data and wrote this manuscript; J.S.F. helped with the harvest module runs; Y.L. greatly improved the experimental design and the manuscript. J.L. supervised the analysis and figure design.

Funding: This research was funded by the National Natural Science Foundation of China (Nos. 31600373 and 41371199) and the K.C. Wong Magna Fund of Ningbo University.

Acknowledgments: We'd like to thank the workgroup from the three Forestry Bureaus for providing details on current forest management and field investigations.

Conflicts of Interest: The authors declare no conflict of interest.

References

1. Bhatti, J.S.; Van Kooten, G.C.; Apps, M.J.; Laird, L.D.; Campbell, I.D.; Campbell, C.; Turetsky, Z.Y.; Banfield, E. Carbon balance and climate change in boreal forests. In *Towards Sustainable Management of the Boreal Forest*; NRC Research Press: Ottawa, ON, Canada, 2003; pp. 799–855.
2. Euskirchen, E.S.; McGuire, A.D.; Chapin, F.S.; Rupp, T.S. The changing effects of Alaska's boreal forests on the climate system. *Can. J. Forest. Res.* **2010**, *40*, 1336–1346. [CrossRef]
3. Liu, Z.; Yang, J.; Chang, Y.; Weisberg, P.J.; He, H.S. Spatial patterns and drivers of fire occurrence and its future trend under climate change in a boreal forest of Northeast China. *Glob. Chang. Biol.* **2012**, *18*, 2041–2056. [CrossRef]
4. Weber, M.G.; Flannigan, M.D. Canadian boreal forest ecosystem structure and function in a changing climate: Impact on fire regimes. *Environ. Rev.* **1997**, *5*, 145–166. [CrossRef]
5. Xu, C.G.; Gertner, G.Z.; Scheller, R.M. Potential effects of interaction between CO_2 and temperature on forest landscape response to global warming. *Glob. Chang. Biol.* **2007**, *13*, 1469–1483. [CrossRef]
6. Kelly, A.E.; Goulden, M.L. Rapid shifts in plant distribution with recent climate change. *Proc. Natl. Acad. Sci. USA* **2008**, *105*, 11823–11826. [CrossRef] [PubMed]
7. Lloyd, A.H.; Bunn, A.G.; Berner, L. A latitudinal gradient in tree growth response to climate warming in the Siberian taiga. *Glob. Chang. Biol.* **2011**, *17*, 1935–1945. [CrossRef]
8. Stueve, K.M.; Isaacs, R.E.; Tyrrell, L.E.; Densmore, R.V. Spatial variability of biotic and abiotic tree establishment constraints across a treelineecotone in the Alaska Range. *Ecology* **2011**, *92*, 496–506. [CrossRef] [PubMed]
9. Seidl, R.; Rammer, W.; Spies, T.A. Disturbance legacies increase the resilience of forest ecosystem structure, composition, and functioning. *Ecol. Appl.* **2014**, *24*, 2063–2077. [CrossRef] [PubMed]
10. Medlyn, B.E.; Duursma, R.A.; Zeppel, M.J.B. Forest productivity under climate change: A checklist for evaluating model studies. *Wires. Clim. Chang.* **2011**, *2*, 332–355. [CrossRef]
11. Johnstone, J.F.; Hollingsworth, T.N.; Chapin, F.S.; Mack, M.C. Changes in fire regime break the legacy lock on successional trajectories in Alaskan boreal forest. *Glob. Chang. Biol.* **2010**, *16*, 1281–1295. [CrossRef]
12. Flannigan, M.; Stocks, B.; Turetsky, M.; Wotton, M. Impacts of climate change on fire activity and fire management in the circumboreal forest. *Glob. Chang. Biol.* **2009**, *15*, 549–560. [CrossRef]
13. Wotton, B.M.; Nock, C.A.; Flannigan, M.D. Forest fire occurrence and climate change in Canada. *Int. J. Wildland Fire* **2010**, *19*, 253–271. [CrossRef]
14. Soja, A.J.; Tchebakova, N.M.; French, N.H.F.; Flannigan, M.D.; Shugart, H.H.; Stocks, B.J.; Sukhinin, A.I.; Parfenova, E.I.; Chapin, F.S.; Stackhouse, J.P.W. Climate-induced boreal forest change: Predictions versus current observations. *Glob. Planet. Chang.* **2007**, *56*, 274–296. [CrossRef]

15. Chapin, F.S.; McGuire, A.D.; Ruess, R.W.; Hollingsworth, T.N.; Mack, M.C.; Johnstone, J.F.; Kasischke, E.S.; Euskirchen, E.S.; Jones, J.B.; Jorgenson, M.T.; et al. Resilience of Alaska's boreal forest to climatic change. *Can. J. For. Res.* **2010**, *40*, 1360–1370. [CrossRef]
16. Holling, C.S. *Engineering Resilience Versus Ecological Resilience*; National Academy Press: Washington, DC, USA, 1996; pp. 31–44.
17. Peterson, G.; Allen, C.R.; Holling, C.S. Ecological resilience, biodiversity, and scale. *Ecosystems* **1998**, *1*, 6–18. [CrossRef]
18. Running, S.W. Is global warming causing more, larger wildfires? *Science* **2006**, *313*, 927–928. [CrossRef] [PubMed]
19. Turner, M.G. Disturbance and landscape dynamics in a changing world. *Ecology* **2010**, *91*, 2833–2849. [CrossRef] [PubMed]
20. Scheffer, M.; Carpenter, S.; Foley, J.A.; Folke, C.; Walker, B. Catastrophic shifts in ecosystems. *Nature* **2001**, *413*, 591–596. [CrossRef] [PubMed]
21. Yao, J.; He, X.Y.; He, H.S.; Chen, W.; Dai, L.M.; Lewis, B.J.; Yu, L.Z. The long-term effects of planting and harvesting on secondary forest dynamics under climate change in northeastern China. *Sci. Rep.* **2016**, *6*, 18490. [CrossRef] [PubMed]
22. He, H.S.; Mladenoff, D.J.; Gustafson, E.J. Study of landscape change under forest harvesting and climate warming-induced fire disturbance. *For. Ecol. Manag.* **2002**, *155*, 257–270. [CrossRef]
23. Gustafson, E.J.; Shvidenko, A.Z.; Sturtevant, B.R.; Scheller, R.M. Predicting global change effects on forest biomass and composition in south-central Siberia. *Ecol. Appl.* **2010**, *20*, 700–715. [CrossRef] [PubMed]
24. He, H.S.; Yang, J.; Shifley, S.R.; Thompson, F.R. Challenges of forest landscape modeling—Simulating large landscapes and validating results. *Landsc. Urban Plan.* **2011**, *100*, 400–402. [CrossRef]
25. Gustafson, E.J.; Shvidenko, A.Z.; Scheller, R.M. Effectiveness of forest management strategies to mitigate effects of global change in south-central Siberia. *Can. J. For. Res.* **2011**, *41*, 1405–1421. [CrossRef]
26. Scheller, R.M.; Mladenoff, D.J. An ecological classification of forest landscape simulation models: Tools and strategies for understanding broad-scale forested ecosystems. *Landsc. Ecol.* **2007**, *22*, 491–505. [CrossRef]
27. Shifley, S.R.; Thompson, F.R.; Dijak, W.D.; Fan, Z.F. Forecasting landscape-scale, cumulative effects of forest management on vegetation and wildlife habitat: A case study of issues, limitations, and opportunities. *For. Ecol. Manag.* **2008**, *254*, 474–483. [CrossRef]
28. Chang, Y.; He, H.S.; Hu, Y.; Bu, R.; Li, X. Historic and current fire regimes in the Great Xing'an Mountains, northeastern China: Implications for long-term forest management. *For. Ecol. Manag.* **2008**, *254*, 445–453. [CrossRef]
29. Xu, H.C. *Forest in Great Xing'an Mountains of China*; Science Press: Beijing, China, 1998.
30. Johnstone, J.F.; Chapin, F.S.; Hollingsworth, T.N.; Mack, M.C.; Romanovsky, V.; Turetsky, M. Fire, climate change, and forest resilience in interior Alaska. *Can. J. For. Res.* **2010**, *40*, 1302–1312. [CrossRef]
31. Van Mantgem, P.J.; Stephenson, N.L.; Byrne, J.C.; Daniels, L.D.; Franklin, J.F.; Fule, P.Z.; Harmon, M.E.; Larson, A.J.; Smith, J.M.; Taylor, A.H.; et al. Widespread increase of tree mortality rates in the western United States. *Science* **2009**, *323*, 521–524. [CrossRef] [PubMed]
32. Wang, W.J.; He, H.S.; Spetich, M.A.; Shifley, S.R.; Thompson, F.R.; Larsen, D.R.; Fraser, J.S.; Yang, J. A large-scale forest landscape model incorporating multi-scale processes and utilizing forest inventory data. *Ecosphere* **2013**, *4*, 1–22. [CrossRef]
33. Parker, G.G.; Harmon, M.E.; Lefsky, M.A.; Chen, J.; Pelt, R.P.; Weiss, S.B.; Thomas, S.C.; Winner, W.E.; Shaw, D.C.; Franklin, J.F. Three-dimensional structure of an old-growth *Pseudotsuga-tsuga* canopy and its implications for radiation balance, and gas exchange. *Ecosystems* **2004**, *7*, 440–453. [CrossRef]
34. Kane, V.R.; Gersonde, R.F.; Lutz, J.A.; McGaughey, R.J.; Bakker, J.D.; Franklin, J.F. Patch dynamics and the development of structural and spatial heterogeneity in Pacific Northwest forests. *Can. J. For. Res.* **2011**, *41*, 2276–2291. [CrossRef]
35. Flato, G.M.; Boer, G.J. Warming asymmetry in climate change simulations. *Geophys. Res. Lett.* **2001**, *28*, 195–198. [CrossRef]
36. Li, X.; He, H.S.; Wu, Z.; Liang, Y.; Schneiderman, J.E. Comparing effects of climate warming, fire, and timber harvesting on a boreal forest landscape in northeastern China. *PLoS ONE* **2013**, *8*, e59747. [CrossRef] [PubMed]

37. Yang, J.; He, H.S.; Gustafson, E.J. A hierarchical fire frequency model to simulate temporal patterns of fire regimes in LANDIS. *Ecol. Model.* **2004**, *180*, 119–133. [CrossRef]
38. Fraser, J.S.; He, H.S.; Shifley, S.R.; Wang, W.J.; Thompson, F.R. Simulating stand-level harvest prescriptions across landscapes: LANDIS PRO harvest module design. *Can. J. For. Res.* **2013**, *43*, 972–978. [CrossRef]
39. Luo, X.; He, H.S.; Liang, Y.; Wang, W.J.; Wu, Z.W.; Fraser, J.B. Spatial simulation of the effect of fire and harvest on aboveground tree biomass in boreal forests of Northeast China. *Landsc. Ecol.* **2014**, *29*, 1187–1200. [CrossRef]
40. Liu, Z.; He, H.S.; Yang, J. Emulating natural fire effects using harvesting in an eastern boreal forest landscape of northeast China. *J. Vege. Sci.* **2012**, *23*, 782–795. [CrossRef]
41. He, H.S.; Mladenoff, D.J.; Crow, T.R. Linking an ecosystem model and a landscape model to study forest species response to climate warming. *Ecol. Model.* **1999**, *114*, 213–233. [CrossRef]
42. Wang, W.; Peng, C.; Kneeshaw, D.D.; Larocque, G.R.; Song, X.; Zhou, X. Quantifying the effects of climate change and harvesting on carbon dynamics of boreal aspen and jack pine forests using the TRIPLEX-Management model. *For. Ecol. Manag.* **2012**, *281*, 152–162. [CrossRef]
43. Bodin, P.; Wiman Bo, L.B. The usefulness of stability concepts in forest management when coping with increasing climate uncertainties. *For. Ecol. Manag.* **2007**, *242*, 541–552. [CrossRef]
44. Perz, S.G.; Muñoz-Carpena, R.; Kiker, G.; Holt, R.D. Evaluating ecological resilience with global sensitivity and uncertainty analysis. *Ecol. Model.* **2013**, *263*, 174–186. [CrossRef]
45. Carpenter, S.R.; Westley, F.; Turner, M.G. Surrogates for resilience of social-ecological systems. *Ecosystems* **2005**, *8*, 941–944. [CrossRef]
46. Folke, C.; Carpenter, S.; Walker, B.; Scheffer, M.; Elmqvist, T.; Gunderson, L.; Holling, C.S. Regime shifts, resilience, and biodiversity in ecosystem management. *Annu. Rev. Ecol. Evol. Syst.* **2004**, *35*, 557–581. [CrossRef]
47. Petchey, O.L.; Gaston, K.J. Effects on ecosystem resilience of biodiversity, extinctions, and the structure of regional species pools. *Theor. Ecol.* **2009**, *2*, 177–187. [CrossRef]
48. Wang, X.P.; Fang, J.Y.; Zhu, B. Forest biomass and root-shoot allocation in northeast China. *For. Ecol. Manag.* **2008**, *255*, 4007–4020. [CrossRef]
49. Zhao, M.; Zhou, G.S. Estimation of biomass and net primary productivity of major planted forests in China based on forest inventory data. *For. Ecol. Manag.* **2005**, *207*, 295–313. [CrossRef]
50. Ivanova, G.A.; Conard, S.G.; Kukavskaya, E.A.; McRae, D.J. Fire impact on carbon storage in light conifer forests of the Lower Angara region, Siberia. *Environ. Res. Lett.* **2011**, *6*, 045203. [CrossRef]
51. Wang, W.J.; He, H.S.; Spetich, M.A.; Shifley, S.R.; Thompson, F.R.; Dijak, W.D.; Wang, Q. A framework for evaluating forest landscape model predictions using empirical data and knowledge. *Environ. Model. Softw.* **2014**, *62*, 230–239. [CrossRef]
52. Wang, X.C.; Zhang, Y.D.; McRae, D.J. Spatial and age-dependent tree-ring growth responses of *Larixgmelinii* to climate in northeastern China. *Trees Struct. Funct.* **2009**, *23*, 875–885. [CrossRef]
53. Luo, X.; Wang, Y.; Zhang, J. Simulating the effects of climate change and fire disturbance on aboveground biomass of boreal forests in the Great Xing'an Mountains, Northeast China. *Chin. J. Appl. Ecol.* **2018**, *29*, 713–724.
54. Leng, W.; He, H.S.; Bu, R.; Dai, L.; Hu, Y.; Wang, X. Predicting the distributions of suitable habitat for three larch species under climate warming in Northeastern China. *For. Ecol. Manag.* **2008**, *254*, 420–428. [CrossRef]
55. Lenoir, J.; Gégout, J.C.; Marquet, P.A.; Ruffray, P.D.; Brisse, H. A significant upward shift in plant species optimum elevation during the 20th century. *Science* **2008**, *320*, 1768–1771. [CrossRef] [PubMed]

Article

Evaluating WorldClim Version 1 (1961–1990) as the Baseline for Sustainable Use of Forest and Environmental Resources in a Changing Climate

Maurizio Marchi [1,*], Iztok Sinjur [2], Michele Bozzano [3] and Marjana Westergren [2]

1 CREA—Research Centre for Forestry and Wood, I-52100 Arezzo, Italy
2 Slovenian Forestry Institute, Vecna pot 2, 1000 Ljubljana, Slovenia; iztok.sinjur@gozdis.si (I.S.); marjana.westergren@gozdis.si (M.W.)
3 European Forest Institute, 53113 Bonn, Germany; michele.bozzano@efi.int
* Correspondence: maurizio.marchi@crea.gov.it; Tel.: +39-0575-353021; Fax: +39-0575-353490

Received: 29 April 2019; Accepted: 27 May 2019; Published: 29 May 2019

Abstract: WorldClim version 1 is a high-resolution, global climate gridded dataset covering 1961–1990; a "normal" climate. It has been widely used for ecological studies thanks to its free availability and global coverage. This study aims to evaluate the quality of WorldClim data by quantifying any discrepancies by comparison with an independent dataset of measured temperature and precipitation records across Europe. BIO1 (mean annual temperature, MAT) and BIO12 (mean total annual precipitation, MAP) were used as proxies to evaluate the spatial accuracy of the WorldClim grids. While good representativeness was detected for MAT, the study demonstrated a bias with respect to MAP. The average difference between WorldClim predictions and climate observations was around +0.2 °C for MAT and −48.7 mm for MAP, with large variability. The regression analysis revealed a good correlation and adequate proportion of explained variance for MAT (adjusted $R^2 = 0.856$) but results for MAP were poor, with just 64% of the variance explained (adjusted $R^2 = 0.642$). Moreover no spatial structure was found across Europe, nor any statistical relationship with elevation, latitude, or longitude, the environmental predictors used to generate climate surfaces. A detectable spatial autocorrelation was only detectable for the two most thoroughly sampled countries (Germany and Sweden). Although further adjustments might be evaluated by means of geostatistical methods (i.e., kriging), the huge environmental variability of the European environment deeply stressed the WorldClim database. Overall, these results show the importance of an adequate spatial structure of meteorological stations as fundamental to improve the reliability of climate surfaces and derived products of the research (i.e., statistical models, future projections).

Keywords: spatial analysis; 1961–1990 normal period; spatial interpolation; geostatistics; ecological mathematics

1. Introduction

Easy access to standardized climate data with global coverage is paramount for the advancement of many ecological studies and to understand future ecosystem services provided by forest systems [1–3] and productive lands in agriculture [4,5]. One of the main aims for researchers dealing with environmental resources has become to forecast possible impacts of climate change on organisms and to evaluate possible mitigation [6–9]. In the past few decades, many conservation strategies have been suggested in order to maintain human well-being and ensure an adequate level of welfare [10] from (relatively) simple management strategies [4,11], including "assisted migration" [12–14], a controversial protocol that includes translocating more adapted or resilient genotypes for conservation or to improve the resilience of ecosystems. Such efforts are often driven by statistical models [14–17] and management simulators [18,19], with both genetic variation and phenotypic plasticity included in the statistical models as covariates [20–22]. However, despite modelling efforts, such studies always require absolutely reliable climate data to be used as both baseline (e.g., 30-year average climate data) and for future predictions. Furthermore, while the uncertainty around GCMs and future trajectories is well known [23–25], information on current ecological limits of forest tree species has also been questioned [26]. In this context, the interest of researchers in gridded climate datasets has grown strongly.

The interpolation method, the spatial resolution and the coverage are the three main features that researchers use to select the most suitable datasets for their research [27–30]. The first release of the WorldClim dataset [31] is probably the most famous gridded climate dataset, widely used for ecological studies and freely available from (www.worldclim.org). Thanks to its high resolution (30 arc-second in the WGS84 reference system and approximately 1 km at the equator), global coverage, and availability, it has been used and cited more than 5200 times since publication [31]. The dataset is suitable for basic and applied studies in ecology, including forestry and ecological modeling [32–34], as well as to construct related datasets such as bio-geographical zones or environmental stratifications [35]. One of the main products of this database is "version 1", representative of the 1961–1990 climate normal period for the whole globe, including Antarctica. This version 1 dataset was generated by interpolating weather station data with the ANUSPLIN software (version 4.3) using latitude, longitude, and elevation as independent variables. The software implements a thin-plate smoothing spline procedure, using every station as a data point. A second-order spline function was fitted by the Authors using the above three variables, which produced the lowest overall cross-validation errors [31]. Considering the ANUSPLIN program creates a continuous surface projection, the LAPGRD program was used to create a global grid of climate surfaces with 30 arc-seconds horizontal and vertical resolution commonly referred to as 1 km^2 resolution. Raster maps for monthly precipitation amount and mean, maximum, and minimum air temperature were then provided. Raw data came from weather stations retrieved from various databases including GHCN, WMO climatological normals, FAOCLIM 2.0, CIAT, and regional databases and, where possible, restricted to the period 1950–2000. Quality control measures were taken to remove duplicate records, giving precedence to the GHCN database. After the quality control check and cleaning, the database consisted of precipitation records from 47,554 locations and mean air temperature from 24,542 locations [31]. Then elevation bias in weather stations was related to latitude and presence of mountain ranges. However, local records from many European countries were not easily accessible and WorldClim climate surfaces for Europe were constructed using 1263 records for air temperature and 2116 for precipitation.

WorldClim version 1 has recently been acknowledged to be representative of the 1961–1990 climate normal period. This time-slice has been widely used as the pre-industrial climate in many papers about the potential impact of climate change on ecosystems [1,3,26,28,36,37] and other ecological fields. Nevertheless, given the detailed description provided by the Authors in their paper, the question remains whether the quality of the WorldClim climate surfaces as a proxy of the climate baseline is adequate in complex environments such as, for instance, the European environment.

The present study aims to assess and quantify the reliability of WorldClim climate raster maps for Europe. We compared WorldClim with observed average values for mean annual temperature and total annual precipitation for the period 1961–1990. Data were retrieved for the whole of Europe building an independent dataset with data from many meteorological services. Then statistical analysis was run in order to evaluate the reliability of this dataset across the study area.

2. Materials and Methods

2.1. Construction and Description of the Database Used for Comparison

To investigate WorldClim's reliability in predicting baseline climate conditions we compiled an independent climate dataset by collecting data from weather services across Europe which were already freely available or delivered upon request (Table 1). All data were specifically requested or downloaded as monthly averaged values over the 30-year normal period (1961–1990). Local monthly air temperature averages (MAT) and precipitation sums (MAP) were aggregated to calculate annual values. In total we retrieved data from 6659 meteostations across Europe, with 1759 records for temperature and 6526 records for precipitation (Figure 1). Most of the records were retrieved for Germany and Sweden with 4825 and 1391 meteostations, respectively, while for some countries, records were much fewer (e.g., Spain, France, Italy) or totally absent (e.g., Serbia, Poland, Romania).

Nevertheless, even if not equally distributed geographically, neither balanced concerning the ecological regions of Europe, we considered the distribution of the collected data as adequate for the purpose. Despite the lack of uniform coverage of both geography and ecological regions, we considered the data collected to be adequate for subsequent analysis.

Moreover we tested the random distribution of MAT and MAP with the randtest package of the R statistical language [38]. The database was carefully checked and cleaned to remove entries with missing data and to geo-reference each record. Very few points (112), corresponding to less than 1% of all the records, lay outside country borders or land masses due to coordinate uncertainties, which reflects the high-quality of the new database. Such records were removed completely from the database in order to avoid any influences on the calculations.

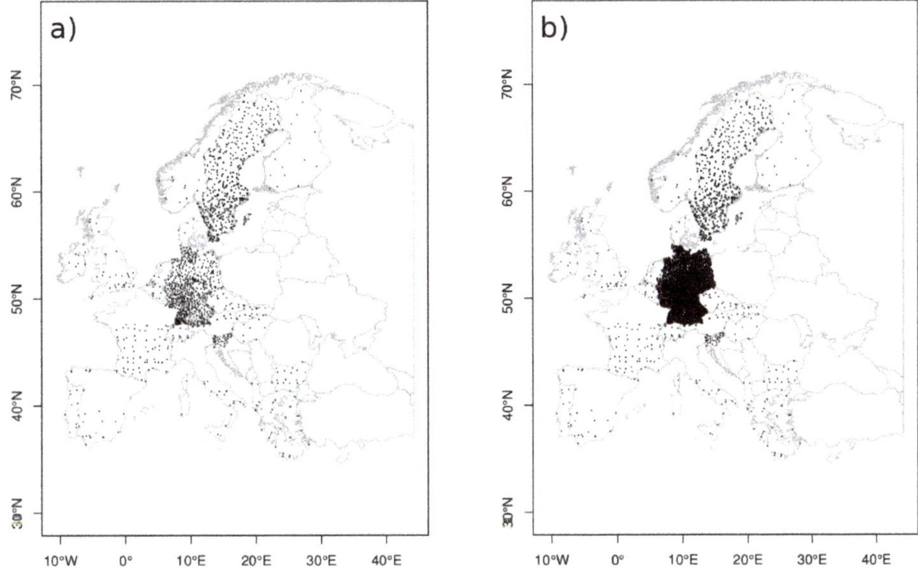

Figure 1. Spatial distribution of the compiled dataset for: (**a**) temperature records and (**b**) precipitation records. Each dot represents a meteorological station. The darker the area, the more data were retrieved.

Table 1. Structure of the compiled database.

Country	Total Meteostations	MAT Records	MAP Records	Data Source
Albania	3	3	3	[39]
Austria	23	21	23	
Belgium	9	9	9	
Bulgaria	18	18	18	
Croatia	1	1	1	
Czech	20	19	20	
Denmark	4	4	4	
Finland	18	17	18	
France	76	75	76	
Germany	4825	719	4733	[40]
Greece	26	25	26	[41]
Hungary	9	9	9	[39]
Ireland	6	6	6	
Italy	30	30	30	
North Macedonia	1	1	1	
Montenegro	1	1	1	
Netherlands	5	5	5	
Norway	18	18	18	
Portugal	18	18	18	
Slovakia	14	14	14	
Slovenia	42	42	42	[42]
Spain	51	51	51	[39]
Sweden	1391	604	1351	[43]
Switzerland	12	11	12	[39]
United Kingdom	38	38	37	
TOTAL	*6659*	*1759*	*6526*	
MEAN	*266*	*70*	*261*	
ST. DEV	*988.64*	*179.54*	*969.08*	

2.2. Comparisons and Statistical Procedures

The BIO1 (mean annual air temperature) and BIO12 (mean total annual precipitation) variables of WorldClim were used as proxies to evaluate the spatial accuracy of raster surfaces. The strata were first downloaded from the official WorldClim web portal. Then, using an overlay function, the corresponding values of the two climate variables were extracted for each meteorological station in our database. A linear regression analysis was then applied to analyze the relationships between the predicted WorldClim value and the observed value in our dataset. The adjusted R^2 was used to measure the amount of environmental variability expressed by WorldClim. Then the difference between the WorldClim value and the observed value (30-years normal value from our database) was calculated for each location of our database. To avoid confusion and mathematical balancing between positive and negative values, which might seriously affect the analysis, both the raw discrepancy (BIAS) and its absolute value (ABIAS) were calculated. To study possible trends across the data, we looked at the relationships between BIAS and the predictors used by the authors of WorldClim during the spatial interpolation process (i.e., latitude, longitude, elevation). Then, we retrieved the complete database of meteorological stations used by the WorldClim authors from www.arcgis.com/home/item.html?id=7644c6e78c1644b4bde2edfc44787520) and clipped to the European environment (Table 2).

We calculated the average distance of each meteorological station in our database from the geographically closest five stations in the WorldClim dataset. We expected a smaller difference where WorldClim stations were denser. Finally, the spatial autocorrelation of BIAS was evaluated using geostatistical analysis implemented in R using the gstat package [44] and modelling the semivariance of BIAS as a function of the spatial distance between records.

The whole structure of the data collection and analysis procedure is graphically reported on Figure 2.

Table 2. Number of meteorological stations per country used by Hijmans et al. [31] in Europe.

Country	Temperature	Precipitation	Country	Temperature	Precipitation
Albania	0	7	Latvia	3	9
Andorra	0	0	Liechtenstein	0	0
Armenia	2	2	Lithuania	16	19
Austria	3	25	Luxembourg	1	6
Belarus	8	22	North Macedonia	7	7
Belgium	3	18	Malta	1	3
Bosnia and Herz.	7	10	Moldova	2	3
Bulgaria	4	15	Monaco	0	0
Croatia	13	13	Montenegro	5	2
Czech Republic	7	16	Netherlands	7	10
Denmark	19	41	Norway	8	54
Estonia	3	12	Poland	18	63
Faeroe Islands	1	1	Portugal	16	18
Finland	19	32	Romania	11	28
France	82	107	Russia	44	124
Georgia	1	20	San Marino	0	0
Germany	89	116	Serbia	23	12
Gibraltar	0	1	Slovakia	3	10
Greece	26	48	Slovenia	6	2
Guernsey	0	0	Spain	60	117
Hungary	8	20	Sweden	16	60
Ireland	16	51	Switzerland	8	20
Isle of Man	0	1	Turkey	513	548
Italy	133	151	Ukraine	22	81
Jersey	0	3	UK	29	188
Summary statistics		**Temperature records**		**Precipitation records**	
TOTAL		1263		2116	
MEAN		25		42	
SD		74.86		84.89	

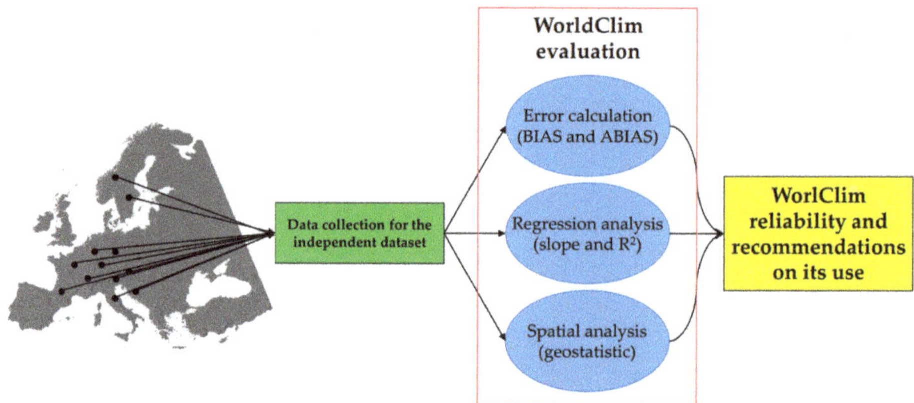

Figure 2. Flowchart of the data collection and statistical analysis we made to test the reliability of WorlClim version 1 data.

3. Results

The compiled database included 25 European countries, albeit with an unbalanced distribution. Overall, an average of 266 records per country (both MAT and MAP) was included in the database. However, the difference among countries was huge, with a standard deviation of ± 988.64 records per country. This large standard deviation was caused by the disproportionate number of records for Sweden and Germany. Temperature (MAT) values ranged from −5.8 °C to 21.2 °C, while precipitation (MAP) was between 104.8 mm and 3318 mm. The mean difference between the interpolated WorldClim values and the observed values was 0.22 °C for temperature and −48.7mm for precipitation (Table 3), with a high coefficient of variation (6.82 for MAT and 3.40 for MAP). BIAS ranged between −10.6 °C and 13.2 °C for MAT and between −1578.1 mm and 950.8 mm for MAP. Mean ABIAS was 0.76 for MAT and 98.56 for MAP.

Results of the regression analysis for MAT and MAP are shown in Figure 3. Residuals of linear models were randomly distributed for both of the analyzed variables and were highly significant ($p < 2.2 \times 10^{-16}$). Concerning MAT, the good correlation and adequate proportion of explained variance point to a low discrepancy between the two datasets; WorldClim explained 86% of the variance (adjusted $R^2 = 0.856$) with a residual random standard error of 1.50°C, intercept of −0.202 °C and slope almost equal to 1 (0.996). The regression line and the expected regression line for a perfect match between the two datasets almost overlapped. For MAP, 64% (adjusted $R^2 = 0.642$) of the variance of the precipitation dataset was explained by a linear regression model, with a residual standard error of 159.6 mm. The match between the two regression lines was considerably low (Figure 3, right) with the slope of the regression coefficient higher than 1. WorldClim was characterized by higher values than observed under 500 mm precipitation and lower values above this threshold. As overall, a general overestimation of MAP values was detected in dry areas (<500 mm) with an underestimation in the remaining zones.

Table 3. Difference between local data and WorldClim's surfaces.

Variable	AVR	SD	CV	MAX	MIN	ABSAVR
MAT [°C]	0.22	1.50	6.82	−10.62	13.21	0.76
MAP [mm]	−48.70	165.35	3.40	−1578.10	950.80	98.56

AVR = average value; SD = standard deviation; CV = coefficient of variation; MAX = maximum difference; MIN = minimum difference; ABSAVR = average of absolute values.

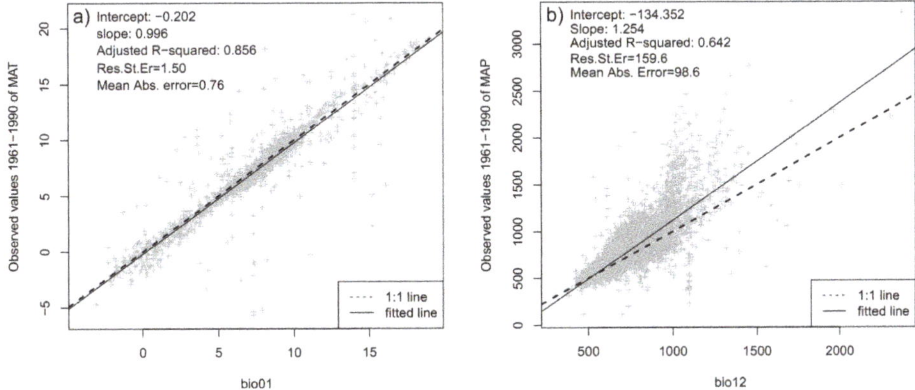

Figure 3. Results of the regression analysis for: (**a**) temperature represented by Bio1 variable of WorldClim database on x-axis and (**b**) precipitation represented by Bio12 variable of WorldClim database versus observed values on the y-axis. Regression coefficients at top-left of each figure. The 1:1 line of a perfect match shown dashed.

No relationship was found between the modelling error (ER) detected for MAT and MAP and the environmental predictors used for spatial interpolation across Europe. Modelled linear regressions explained less than 5% of variation, with one exception (Table 4). This lack of correlation can also be observed in Figure 4 where ABIAS is plotted against the average spatial distance of the "observed" meteorological station from the five WorldClim stations.

Table 4. Linear regression parameters when modelling error (ER) and the environmental predictors. Each predictor was tested separately (ADF5NM=Average distance from the five nearest meteorological stations).

Variable	Predictor	Intercept	Slope	Explained Variance	*p*-Value
MAT	Latitude	0.29	0.000000	0.56%	0.00092
	Longitude	−0.34	0.000000	1.20%	0.00000
	Elevation	0.54	−0.001025	4.95%	0.00000
	ADF5NM	0.09	0.000002	0.18%	0.04138
MAP	Latitude	−45.16	0.000014	0.05%	0.04492
	Longitude	−202.28	0.000061	4.33%	0.00000
	Elevation	8.15	−0.206690	10.26%	0.00000
	ADF5NM	−107.02	0.001059	1.61%	0.00000

The spatial distribution of BIAS in the two most represented countries is shown in Figure 5 for the two investigated variables. Spatial aggregation is especially evident in Sweden, where most of the "large dots" are clustered in the south of the country. For Sweden and Germany, variograms of the MAP variable were fitted by means of an exponential variogram model and revealed a clear spatial autocorrelation, especially for Germany (Figure 6).

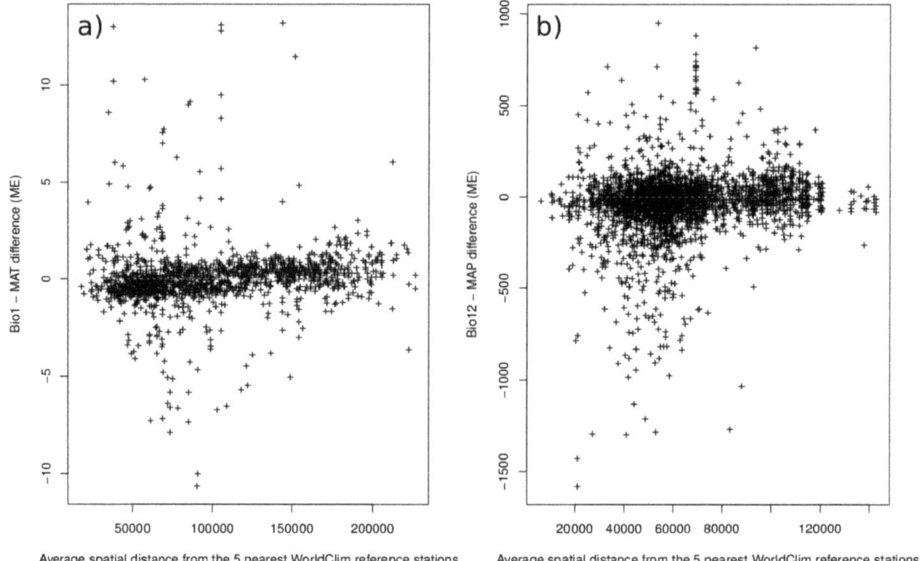

Figure 4. Relationship between the detected differences (WorldClim value, observed) and the average spatial distance of the observed record (new database) from the five nearest WorldClim reference stations for (**a**) temperature and (**b**) precipitation.

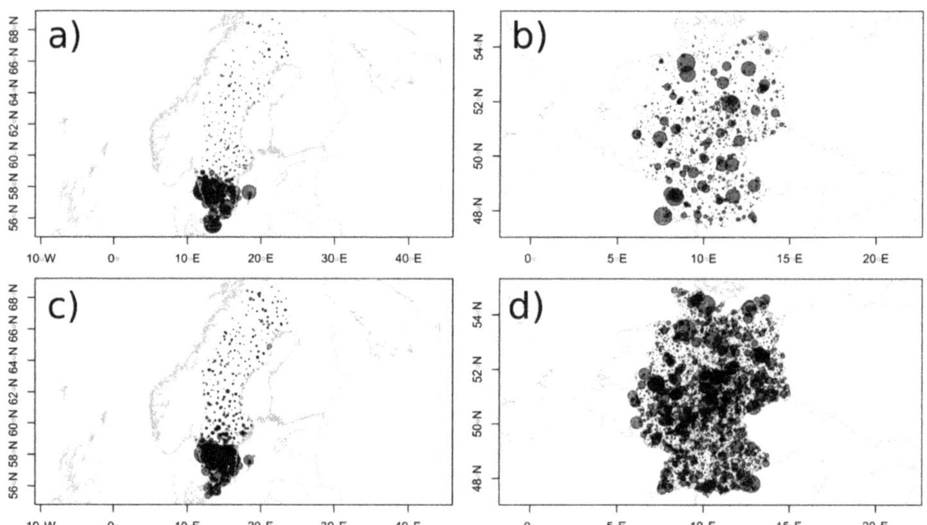

Figure 5. Spatial distribution of ABIAS for temperature (**a** and **b**) and precipitation (**c** and **d**) across Sweden and Germany, the two most sampled countries in the database. The larger the gray dot, the greater the difference between WorldClim and the independent dataset.

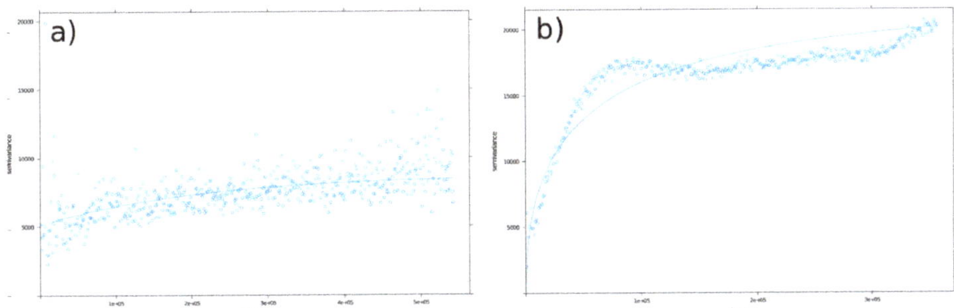

Figure 6. (Semi)variograms for precipitation calculated by the gstat package in R for Sweden (**a**) and Germany (**b**). There is a rather clear spatial structure that might be used as input for further geostatistical procedures (i.e., kriging) in order to adjust WorldClim raster maps.

4. Discussion

The quality of the reference baseline climate has a fundamental role in predictions of the potential impact of climate change on organisms and natural ecosystems. The stability and reliability of the estimated projections calculated from species distribution models [14,45,46], management simulators [18,47], or the estimation of the geographical shift of climate zones [48] rely on the differences between current and future climate. While the representativeness of WorldClim is adequate concerning air temperatures, large differences were found in the precipitation surfaces. Our results demonstrate a systematic difference of 0.76 °C between observed and interpolated values. According to the most recent IPCC report [49], the observed increase in temperature has been around 0.2 °C per decade. As a consequence such difference might affect future projections of WorldClim dataset adding uncertainties on the further modelling efforts [22,50]. In this case a likelihood analysis should be more adequate than deterministic ones and in order to include a sensitivity analysis and evaluate the probability of success of empirical models based on phenotypic plasticity and applied future projections [51]. Precipitation, by contrast, proved to be the main weakness of WorldClim surfaces in Europe. Despite its relatively small spatial extent, the European environment is characterized by many different forest systems that reflect broad climate variability, spanning from the Mediterranean to the Arctic.

The 1961–1990 baseline period is a fundamental dataset for ecological modelling because records from earlier periods were often affected by different instrumentation or changes in observational practice [30,37]. Therefore, numerous studies from climatology to biology, ecology and forestry [36,48,52] have used this baseline period, and WorldClim has been used extensively. We can expect a further warming trend in the next two decades at a rate of about 0.1 °C per decade, due mainly to the slow response of the oceans. As a consequence, even though the linear regression analysis showed a good match between observed and interpolated data (adjusted R^2 = 0.856), the difference is higher than the expected rate of change, which could heavily affect model predictions, adding uncertainties on future projections and smoothing results (i.e., land suitability projections) in an uncontrolled way [14,53–55]. This issue is then amplified when analysing MAP, where higher differences were found in combination with a poor regression analysis result. As a consequence, important biases may be introduced when using WorldClim's precipitation dataset. This is particularly true when WorldClim is used as the reference line and climate projections are locally downscaled and added to the WorldClim surfaces, as in the "Delta method" [56]. As a result, the calculation of climate indices might be difficult. For example, many studies used reference evapotranspiration [57–59] as the main predictor in statistical models [3,60,61]. In this case, the mathematical combination of differences in MAT and MAP might introduce uncontrolled biases through the study area. These biases could represent a critical issue, especially in the Mediterranean and anywhere else that moisture deficit is identified as the most relevant climate driver.

The main advantage of our compiled dataset might be its representativeness at small scale. The authors of WorldClim themselves warn that the high resolution of the climate surfaces does not imply high data quality in all places as this depends on local climate variability, quality and density of observations and the degree of the fitted spline [31]. In a similar study, when compared with PRISM and Daymet datasets for the continental United States, many concerns were expressed, especially regarding the quality of WorldClim's precipitation grids in mountainous areas [34,35,62,63]. For this reason, several studies at regional or national scale at higher resolution (e.g., 100–250 m) preferred the use of meteorological variables obtained at nearby observational sites [28,64–66]. Regardless of the distances of the investigation sites from the locations where meteorological datasets were gathered, orography and land use, and the surrounding area and variable characteristics, must be considered. At small scale their variability may be a strong driver of frequently overlooked heterogeneities, leading to significant discrepancies in transferred datasets used for otherwise appropriate processing methods [67,68].

The lack of any relationship between BIAS and the main physiographic parameters (i.e., latitude, longitude, and elevation) does not allow for any statistical adjustment (e.g., downscaling, locally calibrated lapse rate, etc.) for either temperature or precipitation. However precipitation regimes are very difficult for meteorological stations to record properly and this issue has often been found in other databases [59,69]. Many more data are required, especially in the case of forest monitoring, as a result of the lack of temporal autocorrelation during the timeframe [70].

The need for a freely available and representative global climate dataset is large and growing, as evidenced by WorldClim's citation statistics. These goals can be achieved with local up-to-date monitoring networks, which could play a key role in evaluating global grids at small scale [71] as well as providing data for the construction of additional global climate datasets. Harmonization efforts, as well as increased representativeness of the established networks, are paramount for construction of more accurate climate surfaces. Enhanced data recovery with regular spatial coverage may overcome the lack of dense environmental or climatological sampling [28,70,72,73]. Derived surfaces are fundamental in order to plan future management strategies. For instance, and concerning forestry, additional strata, such as homogeneous climate zones, are needed as a fundamental tool to plan the transfer of genetic resources and reproductive materials across specific geographic areas [74,75]. WorldClim grids were interpolated with spline functions, a fast method known to yield results similar to polynomial functions but without mathematical instability. Such methods do not consider the spatial autocorrelation between observations, only partially achieved by more complex models where latitude and longitude are included as predictive variables [28,76]. Therefore, the exhibited spatial aggregation of the BIAS in the case of denser observations of our dataset (i.e., Sweden and Germany) may be relevant for research activities and improvements of the climate surfaces.

5. Conclusions

A new updated beta version of WorldClim has recently been released for the 1971–2000 time period. This "Version 2" (http://worldclim.org/version2), along with the need for carefully evaluating the quality of records used for modelling and keeping climate databases up-to-date, is an essential requirement for the adequate development of tools and informative systems. The lack of reliability on MAP values can be seen as the main shortcoming of the WorldClim database in Europe and elsewhere. However, precipitation is much more difficult to interpolate, given its low spatial and temporal autocorrelation as well as the lack of statistical relationships with some of the main physiographic parameters, such as elevation. Further research should focus on this parameter, seeking more significant determinants of MAP, given its importance in climate change scenarios where drought stresses are predicted to be the most relevant issue.

Author Contributions: Conceptualization, M.W., M.B. and M.M.; methodology, M.M. and M.W.; software, M.M.; validation, M.M., I.S. and M.W.; formal analysis, M.M.; investigation, M.M., M.B. and M.W.; resources, M.B.; data curation, M.M., I.S. and M.W.; writing—original draft preparation, M.M. and M.W.; writing—review and editing, M.M., I.S., M.B. and M.W.; visualization, M.M.; supervision, M.B. and M.W.; project administration, M.W. and M.B.; funding acquisition, M.M., M.W. and M.B.

Funding: Maurizio Marchi was funded by EU, in the framework of the Horizon 2020 B4EST project "*Adaptive BREEDING for productive, sustainable and resilient FORESTs under climate change*", UE Grant Agreement 773383 (http://b4est.eu/). Marjana Westergren and Iztok Sinjur were funded by the Slovenian Research Agency, Research programme Forest Biology, Ecology and Technology P4-0107.

Acknowledgments: We thank Gregor Vertačnik from National Meteorological Service of Slovenia and Athanasios D. Sarantopoulos from the Hellenic National Meteorological Service Division of Climatology for contributing data.

Conflicts of Interest: The authors declare no conflict of interest.

References

1. Fridman, J.; Holm, S.; Nilsson, M.; Nilsson, P.; Ringvall, A.H.; Ståhl, G. Adapting national forest inventories to changing requirements—The case of the Swedish national forest inventory at the turn of the 20th century. *Silva Fenn.* **2014**, *48*, 1095. [CrossRef]
2. Bussotti, F.; Pollastrini, M. Traditional and novel indicators of climate change impacts on european forest trees. *Forests* **2017**, *8*, 137. [CrossRef]
3. Ray, D.; Petr, M.; Mullett, M.; Bathgate, S.; Marchi, M.; Beauchamp, K. A simulation-based approach to assess forest policy options under biotic and abiotic climate change impacts: A case study on Scotland's National Forest Estate. *For. Policy Econ.* **2019**, *103*, 17–27. [CrossRef]
4. Marchi, M.; Ferrara, C.; Biasi, R.; Salvia, R.; Salvati, L. Agro-forest management and soil degradation in mediterranean environments: towards a strategy for sustainable land use in vineyard and olive cropland. *Sustainability* **2018**, *10*, 2565. [CrossRef]
5. Mouchet, M.A.; Paracchini, M.L.; Schulp, C.J.E.; Stürck, J.; Verkerk, P.J.; Verburg, P.H.; Lavorel, S. Bundles of ecosystem (dis)services and multifunctionality across European landscapes. *Ecol. Indic.* **2017**, *73*, 23–28. [CrossRef]
6. Dyderski, M.K.; Paź, S.; Frelich, L.E.; Jagodziński, A.M. How much does climate change threaten European forest tree species distributions? *Glob. Chang. Biol.* **2018**, *24*, 1150–1163. [CrossRef] [PubMed]
7. Lindner, M.; Maroschek, M.; Netherer, S.; Kremer, A.; Barbati, A.; Garcia-Gonzalo, J.; Seidl, R.; Delzon, S.; Corona, P.; Kolström, M.; et al. Climate change impacts, adaptive capacity, and vulnerability of European forest ecosystems. *For. Ecol. Manag.* **2010**, *259*, 698–709. [CrossRef]
8. Littell, J.S.; McKenzie, D.; Kerns, B.K.; Cushman, S.; Shaw, C.G. Managing uncertainty in climate-driven ecological models to inform adaptation to climate change. *Ecosphere* **2011**, *2*, art102. [CrossRef]
9. Morton, J.F. The impact of climate change on smallholder and subsistence agriculture. *Proc. Natl. Acad. Sci. USA* **2007**, *104*, 19680–19685. [CrossRef]
10. Marchetti, M.; Vizzarri, M.; Lasserre, B.; Sallustio, L.; Tavone, A. Natural capital and bioeconomy: Challenges and opportunities for forestry. *Ann. Silvic. Res.* **2014**, *38*, 62–73.
11. Imeson, A. *Desertification, Land Degradation and Sustainability*; John Wiley & Sons, Inc.: New York, NY, USA, 2012; ISBN 9781119977759.
12. Ferrarini, A.; Selvaggi, A.; Abeli, T.; Alatalo, J.M.; Orsenigo, S.; Gentili, R.; Rossi, G. Planning for assisted colonization of plants in a warming world. *Sci. Rep.* **2016**, *6*, 28542. [CrossRef]
13. Chakraborty, D.; Wang, T.; Andre, K.; Konnert, M.; Lexer, M.J.; Matulla, C.; Schueler, S. Selecting populations for non-analogous climate conditions using universal response functions: The case of douglas-fir in central Europe. *PLoS ONE* **2015**, *10*, e0136357. [CrossRef]
14. Marchi, M.; Ducci, F. Some refinements on species distribution models using tree-level national forest inventories for supporting forest management and marginal forest population detection. *iForest Biogeosci. For.* **2018**, *11*, 291–299. [CrossRef]
15. Millar, C.I.; Stephenson, N.L.; Stephens, S.L. Climate change and forests of the future: Managing in the face of uncertainty. *Ecol. Appl.* **2007**, *17*, 2145–2151. [CrossRef]
16. Fady, B.; Aravanopoulos, F.A.; Alizoti, P.; Mátyás, C.; von Wühlisch, G.; Westergren, M.; Belletti, P.; Cvjetkovic, B.; Ducci, F.; Huber, G.; et al. Evolution-based approach needed for the conservation and silviculture of peripheral forest tree populations. *For. Ecol. Manag.* **2016**, *375*, 66–75. [CrossRef]

17. Zhang, Q.; Wei, H.; Zhao, Z.; Liu, J.; Ran, Q.; Yu, J.; Gu, W.; Zhang, Q.; Wei, H.; Zhao, Z.; et al. Optimization of the fuzzy matter element method for predicting species suitability distribution based on environmental data. *Sustainability* **2018**, *10*, 3444. [CrossRef]
18. Castaldi, C.; Vacchiano, G.; Marchi, M.; Corona, P. Projecting nonnative Douglas fir plantations in southern europe with the forest vegetation simulator. *For. Sci.* **2017**, *63*, 101–110. [CrossRef]
19. Subedi, N.; Sharma, M. Climate-diameter growth relationships of black spruce and jack pine trees in boreal Ontario, Canada. *Glob. Chang. Biol.* **2013**, *19*, 505–516. [CrossRef]
20. Garzón, M.B.; Robson, T.M.; Hampe, A. ΔTraitSDM: Species distribution models that account for local adaptation and phenotypic plasticity. *New Phytol.* **2019**, *222*, 1757–1765. [CrossRef] [PubMed]
21. Valladares, F.; Matesanz, S.; Guilhaumon, F.; Araújo, M.B.; Balaguer, L.; Benito-Garzón, M.; Cornwell, W.; Gianoli, E.; van Kleunen, M.; Naya, D.E.; et al. The effects of phenotypic plasticity and local adaptation on forecasts of species range shifts under climate change. *Ecol. Lett.* **2014**, *17*, 1351–1364. [CrossRef] [PubMed]
22. O'Neill, G.A.; Hamann, A.; Wang, T.L. Accounting for population variation improves estimates of the impact of climate change on species' growth and distribution. *J. Appl. Ecol.* **2008**, *45*, 1040–1049. [CrossRef]
23. Noce, S.; Collalti, A.; Valentini, R.; Santini, M. Hot spot maps of forest presence in the Mediterranean basin. *iForest Biogeosci. For.* **2016**, *9*, 766–774. [CrossRef]
24. Goberville, E.; Beaugrand, G.; Hautekèete, N.C.; Piquot, Y.; Luczak, C. Uncertainties in the projection of species distributions related to general circulation models. *Ecol. Evol.* **2015**, *5*, 1100–1116. [CrossRef] [PubMed]
25. Noce, S.; Collalti, A.; Santini, M. Likelihood of changes in forest species suitability, distribution, and diversity under future climate: The case of Southern Europe. *Ecol. Evol.* **2017**, *7*, 9358–9375. [CrossRef] [PubMed]
26. Pecchi, M.; Marchi, M.; Giannetti, F.; Bernetti, I.; Bindi, M.; Moriondo, M.; Maselli, F.; Fibbi, L.; Corona, P.; Travaglini, D.; et al. Reviewing climatic traits for the main forest tree species in Italy. *iForest Biogeosci. For.* **2019**, *12*, 173–180. [CrossRef]
27. Attorre, F.; Alfo, M.; De Sanctis, M.; Bruno, F. Comparison of interpolation methods for mapping climatic and bioclimatic variables at regional scale. *Int. J. Climatol.* **2007**, *1843*, 1825–1843. [CrossRef]
28. Marchi, M.; Chiavetta, U.; Castaldi, C.; Ducci, F. Does complex always mean powerful? A comparison of eight methods for interpolation of climatic data in Mediterranean area. *Ital. J. Agrometeorol.* **2017**, *1*, 59–72.
29. Azupura, M.; Dos Ramos, K. A comparison of spatial interpolation methods for estimation of average electromagnetic field magnitude. *Prog. Electromagn. Res. M* **2010**, *14*, 135–145. [CrossRef]
30. Mbogga, M.S.; Hamann, A.; Wang, T. Historical and projected climate data for natural resource management in western Canada. *Agric. For. Meteorol.* **2009**, *149*, 881–890. [CrossRef]
31. Hijmans, R.J.; Cameron, S.E.; Parra, J.L.; Jones, G.; Jarvis, A. Very high resolution interpolated climate surfaces for global land areas. *Int. J. Climatol.* **2005**, *25*, 1965–1978. [CrossRef]
32. Molyneux, N.; Soares, I.; Neto, F. Modeling current and future climates using worldclim and diva software: case studies from Timor Leste and India. *J. Crop Improv.* **2014**, *28*, 619–640. [CrossRef]
33. Vacchiano, G.; Motta, R. An improved species distribution model for Scots pine and downy oak under future climate change in the NW Italian Alps. *Ann. For. Sci.* **2015**, *72*, 321–334. [CrossRef]
34. Schueler, S.; Falk, W.; Koskela, J.; Lefèvre, F.; Bozzano, M.; Hubert, J.; Kraigher, H.; Longauer, R.; Olrik, D.C. Vulnerability of dynamic genetic conservation units of forest trees in Europe to climate change. *Glob. Chang. Biol.* **2014**, *20*, 1498–1511. [CrossRef]
35. Metzger, M.J.; Bunce, R.G.H.; Jongman, R.H.G.; Sayre, R.; Trabucco, A.; Zomer, R. A high-resolution bioclimate map of the world: A unifying framework for global biodiversity research and monitoring. *Glob. Ecol. Biogeogr.* **2013**, *22*, 630–638. [CrossRef]
36. Isaac-Renton, M.G.; Roberts, D.R.; Hamann, A.; Spiecker, H. Douglas-fir plantations in Europe: A retrospective test of assisted migration to address climate change. *Glob. Chang. Biol.* **2014**, *20*, 2607–2617. [CrossRef] [PubMed]
37. Wang, T.; Campbell, E.M.; O'Neill, G.A.; Aitken, S.N. Projecting future distributions of ecosystem climate niches: Uncertainties and management applications. *For. Ecol. Manag.* **2012**, *279*, 128–140. [CrossRef]
38. The R Development Core Team. *R: A Language and Environment for Statistical Computing*; R Foundation for Statistical Computing: Vienna, Austria, 2019.
39. KNMI Climate Explorer. Available online: https://climexp.knmi.nl (accessed on 21 April 2016).

40. Deutscher Wetterdienst. Available online: ftp://ftp-cdc.dwd.de/pub/CDC/observations_germany/climate/multi_annual/ (accessed on 13 April 2016).
41. Hellenic National Meteorological Service. Available online: http://www.emy.gr/emy/en/ (accessed on 21 April 2016).
42. National Meteorological Service of Slovenia. Available online: https://meteo.arso.gov.si/met/en/ (accessed on 21 April 2016).
43. Swedish Meteorological and Hydrological Institute. Available online: http://www.smhi.se/klimatdata/meteorologi/temperatur/dataserier-med-normalvarden-1.7354 (accessed on 30 March 2016).
44. Pebesma, E.J. Multivariable geostatistics in S: The gstat package. *Comput. Geosci.* **2004**, *30*, 683–691. [CrossRef]
45. Zhang, L.; Liu, S.; Sun, P.; Wang, T.; Wang, G.; Zhang, X.; Wang, L. Consensus forecasting of species distributions: the effects of niche model performance and niche properties. *PLoS ONE* **2015**, *10*, e0120056. [CrossRef]
46. Hamann, A.; Roberts, D.R.; Barber, Q.E.; Carroll, C.; Nielsen, S.E. Velocity of climate change algorithms for guiding conservation and management. *Glob. Chang. Biol.* **2015**, *21*, 997–1004. [CrossRef]
47. Mairota, P.; Leronni, V.; Xi, W.; Mladenoff, D.J.; Nagendra, H. Using spatial simulations of habitat modification for adaptive management of protected areas: Mediterranean grassland modification by woody plant encroachment. *Environ. Conserv.* **2013**, *41*, 144–156. [CrossRef]
48. Belda, M.; Holtanová, E.; Kalvová, J.; Halenka, T. Global warming-induced changes in climate zones based on CMIP5 projections. *Clim. Res.* **2016**, *71*, 17–31. [CrossRef]
49. IPCC. *Climate Change 2014: Impacts, Adaptation, and Vulnerability. Part A: Global and Sectoral Aspects. Contribution of Working Group II to the Fifth Assessment Report of the Intergovernmental Panel on Climate Change*; Field, C.B., Barros, V.R., Dokken, D.J., Eds.; Cambridge University Press: Cambridge, UK; New York, NY, USA, 2014; p. 1132.
50. Wood, S.N. On confidence intervals for generalized additive models based on penalized regression splines. *Aust. N. Z. J. Stat.* **2006**, *48*, 445–464. [CrossRef]
51. Fréjaville, T.; Fady, B.; Kremer, A.; Ducousso, A.; Garzón, M.B. Inferring phenotypic plasticity and local adaptation to climate across tree species ranges using forest inventory data. *bioRxiv* **2019**, 1–34. [CrossRef]
52. Ray, D.; Bathgate, S.; Moseley, D.; Taylor, P.; Nicoll, B.; Pizzirani, S.; Gardiner, B. Comparing the provision of ecosystem services in plantation forests under alternative climate change adaptation management options in Wales. *Reg. Environ. Chang.* **2015**, *15*, 1501–1513. [CrossRef]
53. Bellucci, A.; Gualdi, S.; Masina, S.; Storto, A.; Scoccimarro, E.; Cagnazzo, C.; Fogli, P.; Manzini, E.; Navarra, A. Decadal climate predictions with a coupled OAGCM initialized with oceanic reanalyses. *Clim. Dyn.* **2013**, *40*, 1483–1497. [CrossRef]
54. Heikkinen, R.K.; Luoto, M.; Araújo, M.B.; Virkkala, R.; Thuiller, W.; Sykes, M.T. Methods and uncertainties in bioclimatic envelope modelling under climate change. *Prog. Phys. Geogr.* **2006**, *30*, 751–777. [CrossRef]
55. Perdinan, P.; Winkler, J.A. Changing human landscapes under a changing climate: Considerations for climate assessments. *Environ. Manag.* **2014**, *53*, 42–54. [CrossRef]
56. Ramirez-Villegas, J.; Jarvis, A. *CIAT Decision and Policy Analysis Working Paper*; CIAT: London, UK, 2010; p. 18.
57. Hargreaves, G.H.; Samani, Z.A. Reference crop evapotranspiration from temperature. *Appl. Eng. Agric.* **1985**, *1*, 96–99. [CrossRef]
58. Maca, P.; Pech, P. Forecasting SPEI and SPI drought indices using the integrated artificial neural networks. *Comput. Intell. Neurosci.* **2016**, *2016*, 3868519. [CrossRef]
59. Vicente-Serrano, S.M.; Beguería, S.; López-Moreno, J.I.; Angulo, M.; El Kenawy, A. A new global 0.5° gridded dataset (1901–2006) of a multiscalar drought index: Comparison with current drought index datasets based on the palmer drought severity index. *J. Hydrometeorol.* **2010**, *11*, 1033–1043. [CrossRef]
60. Svenning, J.C.; Fitzpatrick, M.C.; Normand, S.; Graham, C.H.; Pearman, P.B.; Iverson, L.R.; Skov, F. Geography, topography, and history affect realized-to-potential tree species richness patterns in Europe. *Ecography (Cop.)* **2010**, *33*, 1070–1080. [CrossRef]
61. Colantoni, A.; Ferrara, C.; Perini, L.; Salvati, L. Assessing trends in climate aridity and vulnerability to soil degradation in Italy. *Ecol. Indic.* **2015**, *48*, 599–604. [CrossRef]
62. Hannachi, A.; Jolliffe, I.T.; Stephenson, D.B. Empirical orthogonal functions and related techniques in atmospheric science: A review. *Int. J. Climatol.* **2007**, *27*, 1119–1152. [CrossRef]

63. Daly, C.; Halbleib, M.; Smith, J.I.; Gibson, W.P.; Doggett, M.K.; Taylor, G.H.; Curtis, J.; Pasteris, P.P. Physiographically sensitive mapping of climatological temperature and precipitation across the conterminous United States. *Int. J. Climatol.* **2008**, *28*, 2031–2064. [CrossRef]
64. Blasi, C.; Chirici, G.; Corona, P.; Marchetti, M.; Maselli, F.; Puletti, N. Spazializzazione di dati climatici a livello nazionale tramite modelli regressivi localizzati. *Forest* **2007**, *4*, 213–219. [CrossRef]
65. Wang, T.; Hamann, A.; Spittlehouse, D.; Carroll, C. Locally downscaled and spatially customizable climate data for historical and future periods for North America. *PLoS ONE* **2016**, *11*, e0156720. [CrossRef]
66. Harris, I.; Jones, P.D.; Osborn, T.J.; Lister, D.H. Updated high-resolution grids of monthly climatic observations—The CRU TS3.10 Dataset. *Int. J. Climatol.* **2014**, *34*, 623–642. [CrossRef]
67. Falk, W.; Mellert, K.H. Species distribution models as a tool for forest management planning under climate change: Risk evaluation of Abies Alba in Bavaria. *J. Veg. Sci.* **2011**, *22*, 621–634. [CrossRef]
68. Trivedi, M.R.; Berry, P.M.; Morecroft, M.D.; Dawson, T.P. Spatial scale affects bioclimate model projections of climate change impacts on mountain plants. *Glob. Chang. Biol.* **2008**, *14*, 1089–1103. [CrossRef]
69. Marchi, M.; Castaldi, C.; Merlini, P.; Nocentini, S.; Ducci, F. Stand structure and influence of climate on growth trends of a Marginal forest population of Pinus nigra spp. nigra. *Ann. Silvic. Res.* **2015**, *39*, 100–110.
70. Ferrara, C.; Marchi, M.; Fares, S.; Salvati, L. Sampling strategies for high quality time-series of climatic variables in forest resource assessment. *iForest Biogeosci. For.* **2017**, *10*, 739–745. [CrossRef]
71. Way, R.G.; Oliva, F.; Viau, A.E. Underestimated warming of northern Canada in the Berkeley Earth temperature product. *Int. J. Climatol.* **2017**, *37*, 1746–1757. [CrossRef]
72. Bhowmik, A.K.; Costa, A.C. Representativeness impacts on accuracy and precision of climate spatial interpolation in data-scarce regions. *Meteorol. Appl.* **2014**, *22*, 368–377. [CrossRef]
73. Wong, D.W.; Yuan, L.; Perlin, S. A comparison of spatial interpolation methods for the estimation of air quality data. *J. Expo. Anal. Environ. Epidemiol.* **2004**, *14*, 404–415. [CrossRef] [PubMed]
74. Metzger, M.J.; Brus, D.J.; Bunce, R.G.H.; Carey, P.D.; Gonçalves, J.; Honrado, J.P.; Jongman, R.H.G.; Trabucco, A.; Zomer, R. Environmental stratifications as the basis for national, European and global ecological monitoring. *Ecol. Indic.* **2013**, *33*, 26–35. [CrossRef]
75. De Dato, G.; Teani, A.; Mattioni, C.; Marchi, M.; Monteverdi, M.C.; Ducci, F. Delineation of seed collection zones based on environmental and genetic characteristics for *Quercus suber* L. in Sardinia, Italy. *iForest Biogeosci. For.* **2018**, *11*, 651–659. [CrossRef]
76. Bachmaier, M.; Backes, M. Variogram or Semivariogram? Variance or Semivariance? Allan Variance or Introducing a New Term? *Math. Geosci.* **2011**, *43*, 735–740. [CrossRef]

Article

Biodiversity Observation for Land and Ecosystem Health (BOLEH): A Robust Method to Evaluate the Management Impacts on the Bundle of Carbon and Biodiversity Ecosystem Services in Tropical Production Forests

Kanehiro Kitayama [1,*], Shogoro Fujiki [1], Ryota Aoyagi [1,2], Nobuo Imai [3], John Sugau [4], Jupiri Titin [4], Reuben Nilus [4], Peter Lagan [4], Yoshimi Sawada [1], Robert Ong [4], Frederick Kugan [4] and Sam Mannan [4]

1 Graduate School of Agriculture, Kyoto University, Kitashirakawa Oiwake-cho, Kyoto 606-8502, Japan; fujiki5636@gmail.com (S.F.); aoyagi.ryota@gmail.com (R.A.); sawada.yoshimi.5r@kyoto-u.ac.jp (Y.S.)

2 Smithsonian Tropical Research Institute, Apartado 0843-03092, Balboa, Ancón, Panamá, Panama

3 Department of Forest Science, Tokyo University of Agriculture, Sakuragaoka 1-1-1, Setagaya, Tokyo 156-8502, Japan; i96nobuo@gmail.com

4 Sabah Forestry Department, Locked Bag 68, Sandakan 90009, Malaysia; John.Sugau@sabah.gov.my (J.S.); Jupiri.Titin@sabah.gov.my (J.T.); reuben.nilus@sabah.gov.my (R.N.); peter_lagan@hotmail.com (P.L.); Robert.Ong@sabah.gov.my (R.O.); frederick.kugan@gmail.com (F.K.); Sam.Mannan@sabah.gov.my (S.M.)

* Correspondence: kanehiro@kais.kyoto-u.ac.jp; Tel.: +81-75-753-6080

Received: 31 July 2018; Accepted: 13 November 2018; Published: 15 November 2018

Abstract: The Forest Stewardship Council (FSC) has initiated a new sustainability mechanism, the ecosystem-services certification. In this system, management entities who wish to be certified for the maintenance of ecosystem services (carbon, biodiversity, watershed, soil and recreational services) must verify that their activities have no net negative impacts on selected ecosystem service(s). Developing a robust and cost-effective measurement method is a key challenge for establishing a credible certification system. Using a single method to evaluate a bundle of ecosystem services will be more efficient in terms of transaction costs than using multiple methods. We tested the efficiency of a single method, "biodiversity observation for land and ecosystem health (BOLEH)", to simultaneously evaluate biodiversity and carbon density on a landscape scale in FSC-certified tropical production forests in Sabah, Malaysia. In this method, forest intactness based on the tree-generic compositional similarity with that of a pristine forest was used as an index of biodiversity. We repeated BOLEH in 2009 and 2014 in these forests. Our analysis could detect significant spatiotemporal changes in both carbon and forest intactness during these five years, which reflected past logging intensities and current management regimes in these forests. Enhancement of these ecosystem services occurred in the forest where sustainable management with reduced-impact logging had long been implemented. In this paper, we describe the procedure of the BOLEH method, and results of the pilot test in these forests.

Keywords: Bornean tropical rain forests; ecosystem services enhancement; forest certification; forest intactness; Forest Stewardship Council (FSC); reduced-impact logging; remote sensing; sustainable forest management; tree-community composition

1. Introduction

The Forest Stewardship Council (FSC) forest certification was developed in the early 1990s to reduce the deforestation and forest degradation caused by unsustainable forest management [1].

The FSC forest certification assures buyers/consumers that certified forest products have been produced from forests that have been responsibly/sustainably managed in terms of environment, economy and society. Certified forest products are expected to gain better consumer appeal over uncertified products in the market and eventually drive out uncertified products. Through this market mechanism, the deforestation and forest degradation caused by unsustainable forest management will eventually be mitigated.

The FSC forest certification is issued when management entities comply with a set of criteria for ten principles in their forests and managements. Principle 5 stipulates criteria for enhancing benefits from the forests, including ecosystem services. Based on the FSC forest certification, one assumes that ecosystem services are responsibly managed in certified forests. However, FSC forest certification is by no means a system to quantitatively verify positive management impacts or the enhancement of ecosystem services [1,2]. Based on these considerations, the FSC has increasingly received demands to reliably certify important forest ecosystem services [3–9].

The FSC, therefore, pilot-tested a new system to directly certify the maintenance/enhancement of ecosystem services to promote ecosystem services payments and to further incentivize certificate holders [1,2,5,10–12]. The new system is expected to quantitatively demonstrate for businesses and investors that ecosystem services are maintained/enhanced in certified forests [11]. Ecosystem services include biodiversity conservation, carbon sequestration and storage, watershed services, soil conservation and recreational services [2,12,13]. The FSC suggested generic indicators for each of the five ecosystem services (Annex C, FSC-STD-60-004 V1-0-EN, International Generic Indicators; [13]). However, practical methodologies to evaluate the impacts of forest management on these ecosystem services are still undergoing improvement. Developing robust methodologies to quantitatively evaluate the impacts of forest management on these ecosystem services is urgently needed to support a reliable ecosystem-services certification system [2]. Moreover, many management entities, who wish to apply for the ecosystem-services certification (or other payment for ecosystem services and REDD+ schemes), may lack the technical capacity to measure the maintenance/provision of ecosystem services [1,14–16].

Thus, methodologies to measure the maintenance/provisioning of ecosystem services must be simple and practical enough to be undertaken by management entities on the ground, while scientifically robust enough to accurately measure spatiotemporal changes of each ecosystem service [2]. In view of these challenges, the FSC encourages the development of "a small number of powerful and easy-to-measure proxy impact indicators" [17]. In addition to these technical requirements, the cost of measurements is another issue. Many certification/eco-labelling schemes are already associated with high transaction costs [1,18–20], and further adding a high cost of a ground survey for measuring ecosystem services will discourage stakeholders from applying for the ecosystem-services certification [1].

Jaung et al. [1] suggested bundling ecosystem services to reduce the certification cost per ecosystem service. The certification of multiple ecosystem services also has several other benefits: increasing incomes for management entities, allowing management entities to access diverse ecosystem-services markets, and encouraging management entities to adopt a holistic forest management approach [1]. Certification of bundled ecosystem services can also reduce the potential trade-offs between carbon sequestration and biodiversity conservation [5], which is also an issue in Reducing Emissions from Deforestation and Forest Degradation-plus (REDD+) as a biodiversity safeguard, because enhancing carbon sequestration by industrial plantation of exotic species will sacrifice biodiversity. However, again, the lack of methodologies to measure bundled ecosystem services presents a major challenge [1].

In view of these technical challenges, we report here the results of testing our method to simultaneously measure carbon stock and biodiversity over a large area of tropical production forests. These two ecosystem services, among others, may be relatively easily incorporated into ecosystem-services certification because stakeholder adaptability is high for these services [10].

Our method "biodiversity observation for land and ecosystem health (BOLEH)" was developed primarily for evaluating forest intactness using a metric of tree community composition in logged-over tropical rain forests [21–23]: either species or genus composition could be used in this method. Earlier studies indicated that the use of genus composition was statistically equally reliable as species composition, and yet could reduce identification cost [21–23]. This approach is directly relevant to the FSC ecosystem-services certification because the following generic indicator for biodiversity conservation can be evaluated with our BOLEH:

- Management activities maintain, enhance or restore natural landscape-level characteristics, including forest diversity, composition and structure (stipulated in Principle 5, Annex C; FSC-STD-60-004 V1-0-EN International Generic Indicators; [13]).

Furthermore, the recently published "Ecosystem Services Procedure" [12] suggests "conservation of intact forest landscape" as an important management impact in biodiversity conservation. Our BOLEH method can directly monitor spatiotemporal changes of forest intactness in a landscape context.

In addition, BOLEH can derive carbon stock (density) values using the same dataset as tree community composition. We applied BOLEH to evaluate the spatiotemporal changes of carbon stock and forest intactness (tree community composition) in two contrasting forest management units in 2009 and 2014. Our test sites were Deramakot and Tangkulap in Sabah, Malaysia, which are now certified by the FSC yet having contrasting logging histories [24,25]. Deramakot has been harvested continuously with reduced-impact logging (RIL) since 1995, while Tangkulap was highly degraded by high-impact conventional logging which was operated until 2001. Logging is now fully suspended in Tangkulap, but the recovery of ecosystem services may be slow due to ecological aftereffects of the past high logging impacts and associated collateral damages. We report here the accuracy of estimating carbon stock and forest intactness (or tree community composition as an indicator of biodiversity conservation) using BOLEH and the patterns of spatiotemporal changes of these two ecosystem services between the two contrasting forest management units.

2. Methods

2.1. Site Description

We collected data from the following forest management units (FMUs) in Sabah, Malaysia (Figure 1): Deramakot FMU (5°14′–28′ N, 117°20′–38′ E, 551 km^2) and Tangkulap FMU (5°18′–31′ N, 117°11′–22′ E, 276 km^2). Supplementary data were also collected from Segaliud Lokan FMU (5°20′–27′ N, 117°23′–39′ E, 576 km^2). The three FMUs are adjacent to each other with the same undulating topography and the same equatorial tropical rain forest climate. Selective logging was a major driver of the changes of carbon stock and tree community composition in these forests. The elevation ranges from 12 m to 355 m. The original vegetation is a lowland dipterocarp rainforest throughout the three FMUs.

The areas of the three FMUs were initially logged in 1956, 1970 and 1958, respectively, using conventional logging methods (i.e., high-impact logging with no environmental considerations). In Deramakot, conventional logging continued until 1989, when all logging activities were halted for regrowth [25]. Then, a long-term management plan with RIL was introduced to Deramakot in 1995. RIL is an improved method of selective logging, including pre-harvest inventory, mapping of all canopy trees, directional felling, liana cutting, and planning of skid trails, log decks, and roads [26]. In combination with reduced-impact logging, a longer cutting cycle (i.e., 40 years) was strictly adhered to in accordance with the long-term management plan. These combined approaches helped to preserve forest intactness [24,27]. Deramakot was the first tropical forest certified by the FSC in 1997, and was considered a model of sustainable forest management by the Sabah Government [25,26]. Reflecting this logging history, we observed that the forests inside the Deramakot FMU were less disturbed based on aboveground biomass and various biological communities [27]. In contrast, Tangkulap was repeatedly logged using conventional logging until 2003 and all logging activities were suspended thereafter. Tangkulap was entirely designated as a conservation forest in 2008 and later also certified by the FSC in recognition of its conservation-oriented long-term management plan in 2011. Both Deramakot and Tangkulap are currently managed directly by the Sabah Forestry Department. By contrast, Segaliud Lokan was repeatedly logged using conventional logging until 2002, after which RIL was implemented by the KTS Plantation Corp. Several areas were converted to plantation forests. It is currently certified by the Malaysian Timber Certification Council (MTCC).

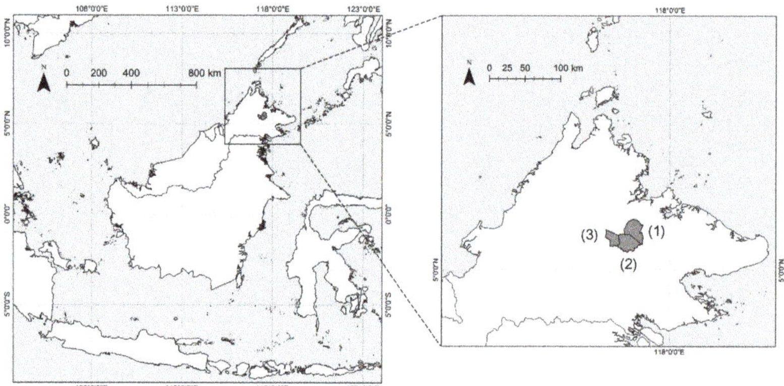

Figure 1. Map indicating the locality of the three forest management units (1, Segaliud Lokan; 2, Deramakot; and 3, Tangkulap), Sabah, Malaysia.

In the following analysis, we compare spatiotemporal changes of ecosystem services, including carbon stock and forest intactness, between Deramakot and Tangkulap. A comparison of these two FMUs may elucidate differences between their respective spatiotemporal patterns in ecosystem services, reflecting the difference in past logging intensity, despite the fact that they are now both FSC-certified. We excluded Segaliud Lokan from this comparison because BOLEH was designed primarily for natural forests. The occurrence of plantation forests in Segaliud Lokan would interfere with the analysis. However, we included natural-forest data of Segaliud Lokan for the following remote-sensing analysis to enhance data points and statistical reliance.

2.2. Field Sampling

A total of 50 circular plots (20-m radius with 0.126 ha area each) were placed in the FMUs from December 2011 until May 2012 (hereafter 2012 inventory). Plots were placed with a stratified random design to represent various forest statuses, ranging from a near pristine high-stock forest to a highly degraded low-stock forest.

Prior to the placement of plots, the entire area was stratified to 6 degradation strata, from primary forest (stratum 1) to highly degraded, open area without trees (stratum 6), with the aid of Landsat Thematic Mapper (TM) band-7 imagery (see Imai et al. [21] for the detailed procedure). Imagery obtained in 2010 was used for the stratification procedure. Briefly, we radiometrically converted Landsat data into Top-of-the-Atmosphere (TOA) reflectance values. Illumination artifacts caused by heterogeneous topography were reduced following the sun-canopy-sensor (SCS) model using TOA values [28]. Shuttle Radar Topography Mission (SRTM) data were employed to correct for illumination artifacts. After masking all non-forest pixels, band-7 values were categorized into 6 strata using a level slicing method. Subsequently, we randomly placed 10 plots in each stratum of the five strata, except for stratum 6. Stratum 6, which is devoid of forests, was not actually sampled. Therefore, we established a total of 50 plots in strata 1 through 5. The distance between any two plots was greater than 100 m in order to minimize spatial autocorrelation.

All trees with $\geq$10 cm in diameter at breast height (dbh) were measured in dbh and identified as genera (actually operationally to species) in each circular plot. Voucher specimens were collected from the trees that could not be identified in the field and later used for identification at the herbarium of the Forest Research Centre, Sandakan. A polyvinyl chloride (PVC) stake was driven into the ground at the center of each circular plot as a permanent marker, and the exact coordinates (latitude and longitude) above the stake were recorded by averaging for two hours using a portable Global Positioning System (GPS).

In August 2014, we repeated the same procedure to inventory the forests as of 2014 (hereafter 2014 inventory). This time, we placed a total of 88 20-m-radius circular plots again evenly in strata 1–5 (actually, some of the plots had a 30 m $\times$ 40 m rectangular shape with approximately the same area as the circular plot because we used these rectangular plots for another purpose). None of the 88 plots overlapped with the 2012 plots because random sampling was deployed. In reality, a total of 50 plots would have been sufficient to derive reliable values of carbon stock and forest intactness according to our standard protocol manuals; however, we added 38 more plots that were placed for another research purpose to enhance model accuracy.

2.3. Pre-Processing of Satellite Images

We used the following two Landsat images to elucidate the temporal changes of carbon and forest intactness between 2009 and 2014 at the landscape level:

- Landsat TM acquired on 11 August 2009 (Path117/Row56)
- Landsat Operational Land Imager (OLI) acquired on 06 June 2014 (Path117/Row56)

The raw digital numbers of each image were converted into top-of-atmosphere radiance. Subsequently, top-of-atmosphere radiance was converted to surface reflectance to compensate for atmospheric scattering and absorption effects using an atmospheric correction algorithm based on the Second Simulation of a Satellite Signal in the Solar Spectrum radiative transfer code (version. 1.1, [29,30]). The topographic effects on illumination were removed using the method described in Ekstrand [31] with Shuttle Radar Topography Mission (SRTM) data. Finally, pixels covered with clouds/shadow were removed according to the procedure described in Fujiki et al. [22]; pixels covered/affected by cloud/shadow were clustered as segments based on a homogeneity criterion with eCognition Developer 8.7 and all segments above a certain threshold were removed as cloud/shadow. Removed segments were filled in using the cloud-free areas of temporally adjoining data. Calibrated cloud-free parts of adjoining secondary images were incorporated into the missing parts of the base image (see Fujiki et al. [22] for the details). We used ERDAS Imagine version.11.0 and ArcGIS 9.3.1 for these pre-processing procedures.

2.4. Estimating Spatiotemporal Changes of Carbon Stock

Above-ground biomass (AGB) of each plot was estimated according to the allometric equation obtained by Chave et al. [32] as:

$$\mathrm{AGB} = \rho \times \exp(-1.499 + 2.148\ln(D) + 0.207(\ln(D))^2 - 0.0281(\ln(D))^3), \quad (1)$$

where D is dbh (cm) and ρ is the wood-specific gravity (g/cm^3). We obtained the wood-specific gravity ρ for the sampled species/genera from various sources (see Imai et al. [21] for the detailed procedure). We estimated AGB of each plot for both the 2012 and 2014 inventories. Subsequently, the estimated AGB of each plot was multiplied by 0.48 to derive the amount of carbon.

A multivariate regression model was established with the amount of carbon per plot as a dependent variable and reflectance of the corresponding pixel on a Landsat image as an independent variable. Here, textural metrics of the 3 $\times$ 3 pixels surrounding each plot were also added as independent variables. The amounts of carbon derived from the 2012 inventory were regressed with the reflectance and textural metrics of the 2009 Landsat imagery (Landsat TM, Path117/Row56, 11 August 2009). The amounts of carbon derived from the 2014 inventory were regressed with the reflectance and textural metrics of the 2014 Landsat imagery (Landsat OLI, Path117/Row56, 06 June 2014). The 2012 inventory data were regressed with the 2009 Landsat imagery without correcting for the tree growth between 2009 and 2012 for a first approximation of the 2009 condition because there were neither 2009 inventory data nor clear 2012 Landsat images; this was logical because all plots measured in 2012 were also present in 2009. However, modeling using the 2012 inventory data without the correction for tree growth would slightly overestimate the modeled 2009 carbon stock.

The following Landsat metrics were used as independent variables: reflectance value of each band (Band1TM/OLI, Band2TM/OLI, Band3TM/OLI, Band4TM/OLI, Band5TM/OLI, Band6OLI, Band7TM/OLI), NDVI [33], normalized difference water index (NDWI) [34,35], normalized difference soil index (NDSI) [36], and enhanced vegetation index (EVI) [37]. We calculated mean value of each Landsat metric within a 20-m radius from the center of a plot to represent the plot. The normalized indices were calculated as follows:

$$\text{NDVI-TM(OLI)} = (\text{band } 4(\text{band5}) - \text{band } 3(\text{band4}))/(\text{band } 4(\text{band5}) + \text{band } 3(\text{band4})), \quad (2)$$

$$\text{NDWI-TM(OLI)} = (\text{band } 3(\text{band4}) - \text{band } 5(\text{band6}))/(\text{band } 3(\text{band4}) + \text{band5}(\text{band6})), \quad (3)$$

$$\text{NDSI-TM(OLI)} = (\text{band } 5(\text{band6}) - \text{band } 4(\text{band5}))/(\text{band } 5(\text{band6}) + \text{band } 4(\text{band5})), \quad (4)$$

$$\text{EVI-TM(OLI)} = 2.5 \times (\text{band } 4(\text{band5}) - \text{band } 3(\text{band4}))/(\text{band } 4(\text{band5}) + 6 \times \text{band3}(\text{band4}) - 7.5 \times \text{band } 1(\text{band2}) + 1), \quad (5)$$

In addition, we used the following metrics as proxies for spectral heterogeneity because degradation of forest canopies might affect the heterogeneity of the spectral pattern: the coefficient of variation (CV), standard deviation (SD), and textures of the gray-level co-occurrence matrix (GLCM) [38]. CV, SD, and textures of the GLCM were derived using a 3 × 3 pixel window based on the reflectance values and each of the normalized indices. GLCM is a tabulation explaining how often different combinations of gray levels occur at a specified distance and orientation in an image object [39]. Homogeneity, contrast, angular second moment, entropy, dissimilarity, correlation, mean, and standard deviation were derived as the indices of the textures. Overall, 120 and 132 metrics were generated based on Landsat TM and OLI, respectively. For developing an adequate regression model, independent variables were selected using a stepwise selection from the full model containing all metrics to avoid multi-collinearity among the independent variables. We did not find significant multi-collinearity because the variance inflation factor (VIF) of selected independent variables, which is an indicator of multi-collinearity, was in all cases less than 10. Established models are shown in Table 1. Subsequently, each of the models (2009 and 2014 models) was extrapolated to the entire area to estimate the amount of carbon outside the inventory plots based on the 2009 or 2014 Landsat imagery.

We took a Monte Carlo approach to show model accuracy and to test for significant differences of mean AGB values between 2009 and 2014 for a given FMU, or between FMUs for a given year (2009 or 2014). Four-fifths of the plots in each stratum (e.g., a total of 40 plots in the case of 2009, and 70 plots in 2014) were randomly selected to construct the AGB models for 2009 and 2014, respectively. The model predictions for AGB of the remaining one-fifth of the plots were regressed with the observed values both for 2009 and 2014, and the adjusted R^2 values were determined. Based on these models, we estimated the mean AGB value each for Deramakot and Tangkulap each in 2009 and 2014. These steps were reiterated 1000 times for 2009 and 2014, respectively, to derive the 95% confidence intervals (CIs) of the adjusted R^2 and the mean AGB values. The statistical tests were conducted using R ver. 3.20 [40].

Table 1. Selected multivariate regression models to explain carbon density based on all plots for 2009 and 2014.

	R^2	Coefficient	SE	T Value	Pr (>|t|)
		2009 (N = 50)			
B7TM		−0.70355	2.70×10^{-4}	9.25227	<0.001
GLCM_mean_NDSI	0.75	0.38178	8.99×10^{-3}	4.97201	<0.001
GLCM_standard_B1TM		−0.15363	1.43×10^{-2}	2.1227	3.92×10^{-2}
		2014 (N = 88)			
B7OLI		−0.61174	3.75×10^{-4}	7.4685	<0.001
NDSI	0.71	−0.32885	1.29	3.94	<0.001
GLCM_correlation_B3OLI		0.00344	2.81×10^{-1}	3.0135	3.44×10^{-3}

N, number of plots used for each model; R^2, adjusted R-squared value; Coefficient, standardized partial regression coefficient; SE, standard error; B1, band 1; B3, band 3; B7 band 7; GLCM, textures of grey-level co-occurrence matrix; NDSI, normalized difference soil index.

2.5. Estimating Spatiotemporal Changes of Forest Intactness

It has been well established that tree-species/genus composition is one of the best indexes of the magnitude of forest degradation [21,23,27]. Similarity of the species/genus composition of a given forest with that of pristine forests decreases linearly with increasing magnitude of degradation of the forest [21,27]. We applied this principle to the forests of Deramakot and Tangkulap and evaluated the status of a given forest in terms of its compositional similarity with pristine forests [21]. The compositional similarity with pristine forests is termed here "forest intactness" and can be mapped through a special extrapolation procedure [22]. Here, we compared generic composition among plots instead of species composition because logging causes similar shifts in generic composition as in species composition [21,23]. The procedure of mapping "forest intactness" for a given time has been described by Fujiki et al. [22]. We elaborate here on the procedure to elucidate temporal changes of "forest intactness".

Firstly, differences in community composition among these plots laid out for estimating carbon storage were examined by using an ordination technique. We used a composite of the data of both the 2012 and 2014 inventories to derive a single data matrix to allow for a comparison between these two years using standardized scores. The Chao distance [41] and the number of trees of each genus were used to calculate a distance matrix. An ordination of plots was conducted with non-metric multidimensional scaling (nMDS) using the "metaMDS" procedure in the Vegan package in R [42].

Although plots were located distantly from each other, we could not completely rule out the possibility of spatial autocorrelations among plots. Therefore, we tested autocorrelations among plots, but did not find significant effects of spatial autocorrelations (see Figure S1, Supplementary Materials).

A multivariate regression model was established with the derived nMDS axis-1 scores of plots as dependent variables, and reflectance and textural metrics of the corresponding pixels on a Landsat image as independent variables; the nMDS axis-1 scores of the 2012 inventory plots were regressed with the 2009 Landsat TM imagery and those from the 2014 inventory were regressed with the 2014 Landsat OLI imagery, as was conducted for the carbon analysis. We did not correct the nMDS axis-1 scores of 2012 when regressed with the 2009 imagery because the tree growth during 3 years between 2009 and 2012 would not substantially change tree-species composition. We used the same Landsat metrics as a carbon stock estimate in this analysis. Established models are indicated in Table 2. Each model was extrapolated to the entire area on the 2009 and 2014 Landsat imagery, respectively. This procedure yielded the maps of forest intactness for 2009 and 2014.

We also took a Monte Carlo approach to test significant differences of mean nMDS axis-1 scores between 2009 and 2014 for a given FMU, or between FMUs for a given year (2009 or 2014) [22]. Four-fifths of the plots in each stratum were randomly selected to construct the 2009 and 2014 model, respectively. Based on these models, we estimated mean nMDS axis-1 scores for Deramakot and for Tangkulap in 2009 and in 2014. The model predictions for nMDS axis-1 scores of the remaining one-fifths of the plots were regressed with the observed values both for 2009 and 2014, and the adjusted R^2 values were collected. Based on these models, we estimated mean nMDS axis-1 scores for Deramakot and for Tangkulap in 2009 and in 2014. These steps were reiterated 1000 times each for 2009 and 2014 to derive the 95% CIs of the adjusted R^2 and the mean reiterated nMDS axis-1 scores. The statistical tests were conducted using R ver. 3.20 [40].

Table 2. Selected multivariate regression models to explain forest intactness (nMDS axis-1 values) based on all plots for 2009 and 2014. Two of the plots for 2014 were not used in the analysis due to incomplete species identification, yielding N = 86.

	R^2	Coefficient	SE	T Value	Pr (>|t|)
Segaliud Lokan–Deramakot–Tangkulap 2009 (N = 50)					
$B5_{TM}$	0.61	−0.53599	3.88×10^{-4}	5.2435	<0.001
GLCM_mean_NDSI		0.37928	1.89×10^{-2}	3.79207	<0.001
Segaliud Lokan-Deramakot-Tangkulap 2014 (N = 86)					
$B7_{OLI}$	0.64	−0.61447	8.92×10^{-4}	6.6404	<0.001
GLCM_mean_EVI		−0.16068	1.65×10^{-2}	2.4093	1.83×10^{-2}
GLCM_homogenity_$B3_{OLI}$		0.16754	7.36	2.5411	1.30×10^{-2}
NDSI		−0.22618	2.99	2.4598	1.60×10^{-2}

N, number of plots used for each model; R^2, adjusted R-squared value; Coefficient, standardized partial regression coefficient; SE, standard error; B3, band 3; B5, band 5; B7 band 7; GLCM, textures of the grey-level co-occurrence matrix; NDSI, normalized difference soil index; EVI, enhanced vegetation index.

3. Results

3.1. Spatiotemporal Changes of Carbon Density

Carbon density values were mainly explained by the short-wave infrared reflectance (band 7), textures of GLCM and NDSI of Landsat metrics based on the stepwise selection (Table 1). A comparison of observed logarithmic vs. predicted logarithmic carbon values is shown in Figure 2; the coefficients of correlations (adjusted R^2) of the full model using all plots were 0.75 for 2009 and 0.71 for 2014. Cross-validation based on 1000 iterations indicated that the 95% CIs of the mean coefficients of correlations were 0.54–0.91 for 2009 and 0.69–0.72 for 2014 (Figure 2).

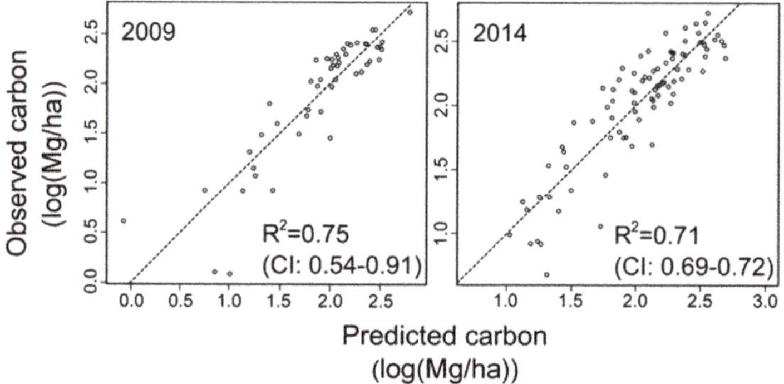

Figure 2. Comparison of observed logarithmic vs. predicted logarithmic carbon values for 2009 (left diagram) and for 2014 (right diagram). Adjusted R^2 values of the full model using all plots were 0.75 for 2009 and 0.71 for 2014. 95% confidence intervals (CIs) of the mean coefficients of correlations were derived from a cross-validation procedure based on 1000 iterations.

Maps indicating the extrapolated carbon densities (Mg/ha) over the entire area of the FMUs for 2009 and 2014 are shown in Figure 3. Mean values (95% CIs) of carbon density (Mg/ha) of Deramakot were 140 (133–151) for 2009 and 170 (162–179) for 2014, while those of Tangkulap were 115 (109–122) for 2009 and 119 (112–126) for 2014 (Figure 3). The mean iterated carbon density value of Deramakot significantly increased from 2009 to 2014 without an overlap in 95% CIs, while that of Tangkulap did not significantly increase and showed a broad overlap of CI (Figure 4).

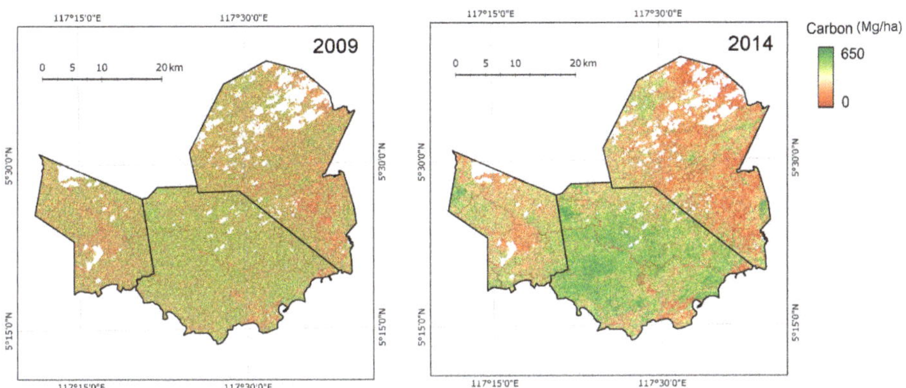

Figure 3. Maps indicating the extrapolated carbon densities (Mg/ha) over the entire area of the three forest management units (FMUs) for 2009 (**left**) and 2014 (**right**).

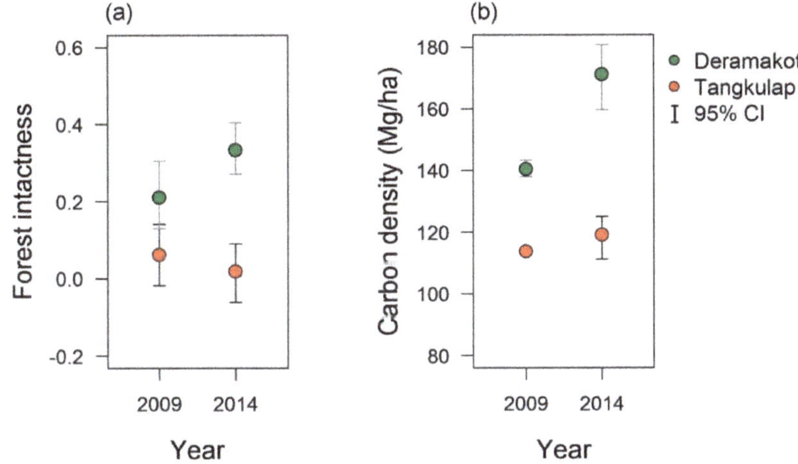

Figure 4. Temporal shifts of the mean (95% CI) forest intactness (**a**) and carbon density (**b**) in Deramakot and Tangkulap FMUs from 2009 to 2014. Green symbols indicate Deramakot and red symbols indicate Tangkulap.

3.2. Spatiotemporal Changes of Forest Intactness

The greatest variance of generic composition occurred along axis 1 of the nMDS ordination (stress values 0.179; Figure S2, Supplementary Materials). Derived nMDS axis-1 scores of plots significantly correlated with logarithmic AGB of the plots (which was considered a surrogate of forest degradation) (adjusted $R^2 = 0.69$, $p < 0.0001$; Figure S3, Supplementary Materials) in line with Imai et al. [21]. Thus, nMDS axis-1 scores of plots were used as a forest "intactness index" in our analysis.

Forest intactness values (nMDS axis-1 scores) based on the stepwise selection were mainly explained by the short-wave infrared reflectance (bands 5 and 7), NDSI, EVI and textures based on the stepwise selection (Table 2). A comparison of observed vs. predicted nMDS axis-1 scores (i.e., intactness index) is shown in Figure 5; coefficients of correlation (R^2) for the full model using all plots were 0.61 for 2009 and 0.64 for 2014. Cross-validation based on 1000 iterations indicated that 95% CIs of the mean coefficients of correlations were 0.19–0.90 for 2009 and 0.42–0.83 for 2014 (Figure 5).

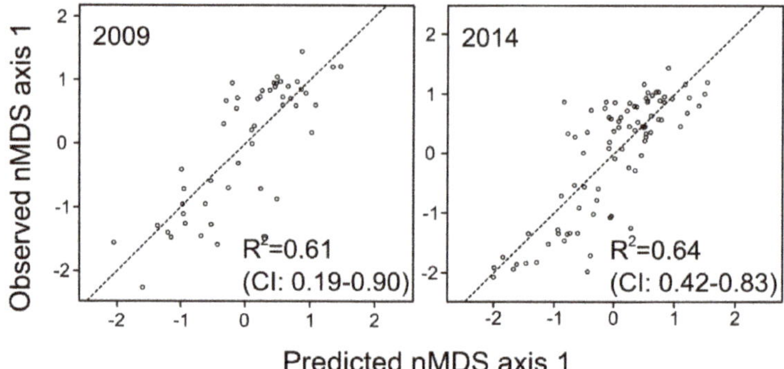

Figure 5. Comparison of observed logarithmic vs. predicted logarithmic nMDS axis-1 values (forest intactness values) for 2009 (left diagram) and for 2014 (right diagram). Adjusted R^2 values of the full model using all plots were 0.61 for 2009 and 0.64 for 2014. 95% CIs of the mean coefficients of correlations were derived from a cross-validation procedure based on 1000 iterations.

Maps indicating the extrapolated forest intactness values (nMDS axis-1 values) over the entire area of the three FMUs are shown for 2009 and 2014 in Figure 6. Mean (95% CI) forest intactness values of Deramakot were 0.211 (0.130–0.306) for 2009 and 0.333 (0.271–0.405) for 2014, while those of Tangkulap were 0.062 (−0.018–0.142) for 2009 and 0.019 (−0.061–0.091) for 2014 (Figure 6). The mean iterated forest intactness value of Deramakot increased with a slight overlap in 95% CIs, while that of Tangkulap did not significantly increase and showed a broad overlap of CI (Figure 4).

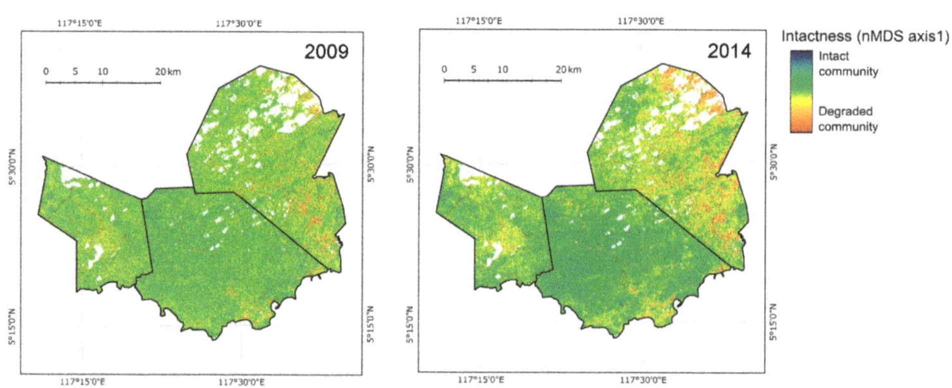

Figure 6. Maps indicating the extrapolated forest intactness (nMDS axis-1 scores) over the entire area of the three FMUs for 2009 (**left**) and 2014 (**right**).

4. Discussion

The BOLEH method successfully elucidated the spatiotemporal changes of carbon density and forest intactness as maps across the three FMUs between 2009 and 2014. The procedure to map forest intactness was already well described by Fujiki et al. [22], who demonstrated forest intactness maps of six FMUs across Borneo. They demonstrated that the mean values of and the frequency distributions of forest intactness significantly differed among FMUs, reflecting the forest management schemes. Our current study further demonstrates that the BOLEH method is useful to monitor temporal changes of forest intactness in a given FMU. Particularly, our BOLEH method could reveal a significant increase of mean carbon density value and a marginally significant increase of mean forest intactness value in Deramakot between August 2009 and June 2014 (time of acquisition of Landsat data) in spite of the continued production of timber. Deramakot FMU produced a total of 43,023 m^3 volume of round logs in two compartments during the corresponding five years. The increases in both mean carbon density and forest intactness suggest that the regrowth in fallow compartments outweighed the harvest. Our BOLEH method based on landscape analysis could successfully evaluate such a balance between negative impacts of harvesting and positive impacts of leaving areas fallow.

Co-benefits of sustainable forestry with RIL on the carbon storage function have already been suggested by Imai et al. [24], Langner et al. [43] and Langner et al. [44], who compared Deramakot with Tangkulap with a snapshot of carbon density in a given year using a space-for-time approach. In our current analysis, we have directly proved the co-benefits of sustainable forestry and RIL on the carbon storage and forest intactness by continuous monitoring. The vast majority of production forests across Malaysia and Indonesia in Borneo have been mildly to highly degraded due to past multiple entries of logging. Deramakot is not an exception. The mean annual harvest with RIL from such degraded secondary forests is 30 m^3/ha and the mean annual harvest area is 347 ha over the 20 years between 1995 and 2016 in Deramakot (unpublished statistical data). Collateral damages are also unavoidable even if timber is carefully harvested under RIL. If we assume that harvest practices with collateral damage produce 30 m^3/ha of waste (i.e., equivalent to the harvested volume), a total of 60 m^3/ha of trees are removed annually from 347 ha, giving rise to a total removal of 20,820 m^3/year (or 10,410 tons carbon/year in Deramakot assuming that 1 m^3 volume is equivalent to 0.5 ton carbon). On the other hand, there are a total of approximately 54,653 ha of fallow compartments each year in Deramakot. Tree regrowth with an increment of merely 0.19 ton carbon/ha/year will make up for the removed carbon of 10,410 tons. Our results indicate that mean carbon density significantly increased from 140 (133–151) ton/ha to 170 (162–179) ton/ha during 5 years from 2009 to 2014, which is equivalent to 6 tons carbon/ha/year including harvested compartments. The ratio of mean harvest area to the fallow area is 347 ha to 54,653 ha, and this wide ratio (i.e., a long rotation) is important to sustain the surplus in carbon budget in secondary production forests. Therefore, Bornean tropical production forests with a comparable biomass stock and a comparable management plan (a long rotation period and moderate harvesting) to those of Deramakot will likely be assured of an increase of carbon stock.

On the other hand, mean forest intactness only marginally increased between 2009 and 2014 in Deramakot, as indicated by the fact that the 95% CIs overlapped slightly between these years. Why did the mean forest intactness not significantly increase while the mean carbon density significantly increased during the same period? Probably, a longer time is required for tree communities to recover in species/genus composition, while carbon increments can occur as a simple function of time. There must be a shift of tree communities (from pioneer to climax species) in order to demonstrate a significant increase of forest intactness. For instance, a vast area of young pioneer -tree stands will support a rapid carbon increment but not an increment in forest intactness because species composition is rather stabilized in such stands. Therefore, monitoring of both ecosystem services (carbon and forest intactness) is necessary for the meaningful evaluation of ecosystem integrity/health.

Why the mean carbon density and forest intactness did not increase in Tangkulap between 2009 and 2014 in spite of the suspension of logging operations is an intriguing question. As reported by Kitayama [27], the natural forests of Tangkulap have been highly degraded by past repeated logging;

they are currently dominated by stands of pioneer trees such as *Macaranga* or by fern grasslands. The shift from such pioneer stands to climax stands will be extremely slow. On the other hand, the slow recovery of carbon stock is puzzling because the building phase of secondary succession is known to accumulate carbon at a rapid rate. Poorter et al. [45] reported a median carbon accumulation rate of 3.05 ton-C/ha/year in secondary forests at 20-year age in the Neotropics. Our estimate of the carbon accumulation rate in Tangkulap is merely 0.8 ton-C/ha/year, although this value was not statistically significant. Probably, the occurrence of vast, thick fern stands is related to the slow recovery because the regeneration in such thick fern stands is extremely slow [46]. When forest regrowth is analyzed, such fern stands tend to be avoided for sampling by ecologists. Moreover, mean tree mortality may be greater in senescent pioneer-tree stands (Imai, personal observation). Therefore, landscape-level evaluations based on remotely sensed data like our BOLEH are required to elucidate the spatially and temporally explicit patterns of carbon (and forest intactness). Another possible reason for the slow carbon accumulation in Tangkulap could be the use of the 2012 inventory data for developing the 2009 carbon model without correcting for the tree growth between 2009 and 2012, because this would slightly overestimate the carbon density for 2009. However, this would not be an important weakness of our study, because our major objectives here were to test our algorithms, but not to investigate the carbon dynamics *per se*.

The FSC ecosystem-services certification has set forth a standard by which forest managers must verify at least the non-existence of net negative management impacts on ecosystem services [12]. The 5-year period seems to be adequate for verifying that there are no net negative management impacts on carbon density or forest intactness using our BOLEH in the case of Deramakot. However, in an FMU where the harvested volume outweighs regrowth (i.e., net negative impacts), the 5-year period may not be long enough to demonstrate a statistically significant negative impact (i.e., reduction in mean values of carbon and forest intactness) because 95% CIs tend to be fairly wide in our method. Fujiki et al. [22] discussed the reasons for wide CIs and suggested that the number of plots was too small for cross-validation. When outlier plots are used to develop a model (to explain carbon or forest intactness with Landsat metrics), residuals become large, giving rise to disproportionately large or small mean values; this must be the case for the very low coefficient of correlation (i.e., 0.19) in the 2009 nMDS cross-validation. The coefficient of correlation for the lower 95% CI in the 2014 nMDS cross-validation was relatively high (i.e., 0.42) probably because we used a total of 86 plots. Here, we still suggest using a total of 50 plots as standard sampling for reducing field efforts. If we maintain this standard sampling procedure, wide 95% CIs will be inherent to our methods. Therefore, forest managers should use our method only when mean values of carbon and forest intactness tend to increase or to be stable in order to verify the non-existence of net negative impacts for the FSC ecosystem-services certification. The 5-year period is equivalent to one period of an FSC forest certification. If BOLEH is incorporated into the assessment procedures of the regular FSC certification for sustainable management, forest managers can verify the enhancement of ecosystem services with low cost in addition to criteria pertinent to environmental values and impacts (principle 6) and monitoring and assessment (principle 8). It should be noted, however, that forest intactness cannot be used as a surrogate of richness of biological taxa because compositional distances from a pristine forest are unrelated to richness of biological taxa [21,27].

5. Conclusions

The BOLEH method could elucidate spatiotemporal changes both in carbon stock and biodiversity (tree community composition or forest intactness) during 5 years for nearly 100,000 ha of logged-over tropical rain forests, which reflected past logging intensities and current management regimes. Field sampling can be completed within a few months by a team of four or five workers with minimal support from experts. In our pilot test, expertise of tree identification was provided by the local Forest Research Centre. When such expertise is not locally available, tree identification must rely on external botanical experts. However, the BOLEH method uses genus instead of species abundance, which can reduce the burden of tree identification. The BOLEH method is useful as a standard method to verify the maintenance/enhancement of carbon stock and biodiversity conservation to meet the requirements of the FSC ecosystem-services procedure.

The whole of the above procedures from field sampling to statistical analyses to extrapolation are compiled as a protocol manual and can be downloaded at the following website:

- http://www.rfecol.kais.kyoto-u.ac.jp/files/Boleh%20manual%202017.1.zip

Supplementary Materials: The following are available online at http://www.mdpi.com/2071-1050/10/11/4224/s1, Figure S1: Effects of distance on the difference in the nMDS axis-1 scores of given paired vegetation plots, Figure S2: Results of the nMDS analysis, Figure S3: Correlation of nMDS axis-1 values with logarithmic above-ground biomass.

Author Contributions: Conceptualization, K.K., F.K. and S.M.; Data curation, S.F., R.A., N.I., J.S., J.T., R.N., P.L. and R.O.; Formal analysis, S.F., R.A. and N.I.; Funding acquisition, K.K.; Investigation, K.K., S.F., R.A., N.I., J.S., J.T., R.N., P.L. and R.O.; Methodology, K.K., S.F., R.A. and N.I.; Project administration, K.K. and Y.S.; Resources, J.S., J.T., R.N., P.L., R.O., F.K. and S.M.; Supervision, K.K., F.K. and S.M.; Validation, S.F. and R.A.; Visualization, S.F. and Y.S.; Writing–original draft, K.K., S.F. and Y.S.; Writing–review & editing, R.A., N.I., J.S., J.T., R.N., P.L., R.O., F.K. and S.M.

Funding: This research was funded by the Global Environmental Research Fund D-1006 and 1-1403 of the Ministry of the Environment, Japan to K.K., and by the United Nations University GGS Project Fund to K.K.

Acknowledgments: This is a contribution to the FSC ecosystem services procedure.

Conflicts of Interest: The authors declare no conflicts of interest.

References

1. Jaung, W.; Putzel, L.; Bull, G.Q.; Guariguata, M.R.; Sumaila, U.R. Estimating demand for certification of forest ecosystem services: A choice experiment with Forest Stewardship Council certificate holders. *Ecosyst. Serv.* **2016**, *22*, 193–201. [CrossRef]
2. Meijaard, E.; Sheil, D.; Guariguata, M.R.; Nasi, R.; Sunderland, T.C.H.; Putzel, L. *Ecosystem Services Certification: Opportunities and Constrains*; CIFOR: Bogor, Indonesia, 2011. [CrossRef]
3. Bass, S.; Simula, M. *Independent Certification/Verification of Forest Management*; Background Paper; The World Bank/WWF Alliance Workshop: Washington, DC, USA, 1999.
4. Griscom, B.; Ellis, P.; Francis, E.P. Carbon emissions performance of commercial logging in East Kalimantan, Indonesia. *Glob. Chang. Biol.* **2014**, *20*, 923–937. [CrossRef] [PubMed]
5. Jaung, W.; Bull, G.Q.; Putzel, L.; Kozak, R.; Elliott, C. Bundling forest ecosystem services for FSC certification: An analysis of stakeholder analysis. *Int. For. Rev.* **2016**, *18*, 452–465. [CrossRef]
6. Merger, E.; Dutschke, M.; Verchot, L. Options for REDD+ voluntary certification to ensure net GHG benefits, poverty alleviation, sustainable management of forests and biodiversity conservation. *Forests* **2011**, *2*, 550–577. [CrossRef]
7. Pettenella, D.; Brotto, L. Governance features for successful REDD+ projects organization. *For. Policy Econ.* **2012**, *18*, 46–52. [CrossRef]
8. Rametsteiner, E.; Simula, M. Forest certification—An instrument to promote sustainable forest management? *J. Environ. Manag.* **2003**, *67*, 87–98. [CrossRef]
9. Vogt, K.A.; Larson, B.C.; Gordon, J.C.; Vogt, D.J.; Fanzeres, A. *Forest Certification: Roots, Issues Challenges and Benefits*; Yale University, CRC Press: Washington, DC, USA, 2000; p. 374. ISBN 9780849315855.
10. Jaung, W.; Putzel, L.; Bull, G.Q.; Kozak, R.; Elliott, C. Forest Stewardship Council certification for forest ecosystem services: An analysis of stakeholder adaptability. *For. Policy Econ.* **2016**, *70*, 91–98. [CrossRef]
11. Savilaakso, S.; Guariguata, M.R. Challenges for developing Forest Stewardship Council certification for ecosystem services: How to enhance local adoption? *Ecosyst. Serv.* **2017**, *28*, 55–66. [CrossRef]
12. FSC. *Ecosystem Services Procedure: Impact Demonstration and Market Tools (FSC-PRO-30-006 V1-0 EN)*; FSC: Bonn, Germany, 2018; Available online: https://ic.fsc.org/en/document-center/id/328 (accessed on 10 April 2017).
13. FSC. *International Genetic Indicators (FSC-STD-60-004 V1-0 EN)*; FSC: Bonn, Germany, 2015; Available online: https://ic.fsc.org/en/document-center/id/87 (accessed on 10 April 2017).

14. Pagiola, S.; Arcenas, A.; Platais, G. Can payments for environmental services help reduce poverty? An exploration of the issues and the evidence to date from Latin America. *World Dev.* **2005**, *33*, 237–253. [CrossRef]
15. Romijn, A.; Herold, M.; Kooistra, L.; Murdiyarso, D.; Verchot, L. Assessing capacities of non-Annex I countries for national forest monitoring in the context of REDD+. *Environ. Sci. Policy* **2012**, *19–20*, 33–48. [CrossRef]
16. Wunder, S.; Engel, S.; Pagiola, S. Taking Stock: A Comparative Analysis of Payments for Environmental Services Programs in Developed and Developing Countries. *Ecol. Econ.* **2008**, *65*, 834–852. [CrossRef]
17. FSC. *Demonstrating the Impact of Forest Stewardship on Ecosystem Servies (FSC-PRO-30-006)*; FSC: Bonn, Germany, 2015. Available online: https://ic.fsc.org/en/fsc-system/current-processes/fsc-pro-30-006 (accessed on 10 April 2017).
18. Ferraro, P.J. Asymmetric information and contract design for payments for environmental services. *Ecol. Econ.* **2008**, *65*, 810–821. [CrossRef]
19. Muradian, R.; Corbera, E.; Pascual, U.; Kosoy, N.; May, P. Reconciling theory and practice: An alternative conceptual framework for understanding payments for ecosystem services. *Ecol. Econ.* **2010**, *69*, 1202–1208. [CrossRef]
20. Muradian, R.; Rival, L. Between markets and hierarchies: The challenge of governing ecosystem services. *Ecosyst. Serv.* **2012**, *1*, 93–100. [CrossRef]
21. Imai, N.; Tanaka, A.; Samejima, H.; Sugau, J.B.; Pereira, J.T.; Titin, J.; Kurniawan, Y.; Kitayama, K. Tree community composition as an indicator in biodiversity monitoring of REDD+. *For. Ecol. Manag.* **2014**, *313*, 169–179. [CrossRef]
22. Fujiki, S.; Aoyagi, R.; Tanaka, A.; Imai, N.; Kusma, A.D.; Kurniawan, Y.; Lee, Y.F.; Sugau, J.B.; Pereira, J.T.; Samejima, H.; et al. Large-scale mapping of tree-community composition as a surrogate of forest degradation in Bornean tropical rain forests. *Land* **2016**, *5*, 45. [CrossRef]
23. Aoyagi, R.; Imai, N.; Fujiki, S.; Sugau, J.B.; Pereira, J.T.; Kitayama, K. The mixing ratio of tree functional groups as a new index for biodiversity monitoring in Bornean production forests. *For. Ecol. Manag.* **2017**, *403*, 27–43. [CrossRef]
24. Imai, N.; Samejima, H.; Langner, A.; Ong, R.C.; Kita, S.; Titin, J.; Chung, A.Y.C.; Lagan, P.; Lee, Y.F.; Kitayama, K. Co-benefits of sustainable forest management in biodiversity conservation and carbon sequestration. *PLoS ONE* **2009**, *4*, e8267. [CrossRef] [PubMed]
25. Ong, R.C.; Langner, A.; Imai, N.; Kitayama, K. Management history of the study sites: The Deramakot and Tangkulap forest reserves. In *Co-Benefits of Sustainable Forestry: Ecological Studies of a Certified Bornean Rain Forest*; Kitayama, K., Ed.; Springer: Tokyo, Japan, 2013; pp. 1–22. ISBN 9784431541400.
26. Lagan, P.; Mannan, S.; Matsubayashi, H. Sustainable use of tropical forests by reduced-impact logging in Deramakot Forest Reserve, Sabah, Malaysia. *Ecol. Res.* **2007**, *22*, 414–421. [CrossRef]
27. Kitayama, K. *Co-Benefits of Sustainable Forestry: Ecological Studies of a Certified Bornean Rain Forest*; Springer: Tokyo, Japan, 2013; ISBN 9784431541400.
28. Gu, D.; Gillespie, A. Topographic normalization of Landsat TM images of forest based on subpixel sun–canopy–sensor geometry. *Remote Sens. Environ.* **1998**, *64*, 166–175. [CrossRef]
29. Kotchenova, S.Y.; Vermote, E.F.; Matarrese, R.; Klemm, F.J., Jr. Validation of a vector version of the 6S radiative transfer code for atmospheric correction of satellite data. Part I: Path radiance. *Appl. Opt.* **2006**, *45*, 6762–6774. [CrossRef] [PubMed]
30. Vermote, E.F.; El Saleous, N.; Justice, C.O.; Kaufman, Y.J.; Privette, J.L.; Remer, L.; Roger, J.C.; Tanre, D. Atmospheric correction of visible to middle-infrared EOS-MODIS data over land surfaces: Background, operational algorithm and validation. *J. Geophys. Res. Atmos.* **1997**, *102*, 17131–17141. [CrossRef]
31. Ekstrand, S. Landsat TM-based forest damage assessment: Correction for topographic effects. *Photogramm. Eng. Remote Sens.* **1996**, *62*, 151–161.
32. Chave, J.; Andalo, C.; Brown, S.; Cairns, M.A.; Chambers, J.Q.; Eamus, D.; Fölster, H.; Fromard, F.; Higuchi, N.; Kira, T.; et al. Tree allometry and improved estimation of carbon stocks and balance in tropical forests. *Oecologia* **2005**, *145*, 87–99. [CrossRef] [PubMed]
33. Rouse, J.W., Jr.; Haas, R.H.; Schell, J.A.; Deering, D.W.; Harlan, J.C. *Monitoring the Vernal Advancement and Retrogradation (Greenwave Effect) of Natural Vegetation*; NASA/GSFC Type III final report; Remote Sensing Center, Texas A&M University: College Station, TX, USA, 1974; p. 371.

34. Gao, B.C. NDWI—A normalized difference water index for remote sensing of vegetation liquid water from space. *Remote. Sens. Environ.* **1996**, *58*, 257–266. [CrossRef]
35. McFeeters, S.K. The use of the Normalized Difference Water Index (NDWI) in the delineation of open water features. *Int. J. Remote Sens.* **1996**, *17*, 1425–1432. [CrossRef]
36. Takeuchi, W.; Yasuoka, Y. Development of normalized vegetation, soil and water indices derived from satellite remote sensing data. *J. Jpn. Soc. Photogramm.* **2004**, *43*, 7–19. [CrossRef]
37. Huete, A.; Didan, K.; Miura, T.; Rodriguez, E.P.; Gao, X.; Ferreira, L.G. Overview of the radiometric and biophysical performance of the MODIS vegetation indices. *Remote Sens. Environ.* **2002**, *83*, 195–213. [CrossRef]
38. Haralick, R.M. Statistical image texture analysis. In *Handbook of Pattern Recognition and Image Processing*; Young, T.Y., Fu, K.S., Eds.; Academic Press: Orland, FL, USA, 1986; Volume 68, pp. 247–279.
39. Baatz, M.; Benz, U.; Dehghani, S.; Heynen, M.; Höltje, A.; Hofmann, P.; Lingenfelder, I.; Mimler, M.; Sohlbach, M.; Weber, M. *eCognition Professional User Guide 4*; Definiens Imaging: Munich, Germany, 2004.
40. R Core Team. *A Language and Environment for Statistical Computing*; R Foundation for Statistical Computing: Vienna, Australia, 2014.
41. Chao, A.; Chazdon, R.L.; Colwell, R.K.; Shen, T.-J. A new statistical approach for assessing similarity of species composition with incidence and abundance data. *Ecol. Lett.* **2005**, *8*, 148–159. [CrossRef]
42. Oksanen, J.; Blanchet, F.G.; Kindt, R.; Legendre, P.; Minchin, P.R.; O'Hara, R.; Simpson, G.L.; Solymos, P.; Stevens, M.H.H.; Wagner, H. Package 'Vegan'. Community Ecology Package Version 2.0-9. 2013. Available online: http://CRAN.R-project.org/package=vegan (accessed on 20 September 2013).
43. Langner, A.; Samejima, H.; Ong, R.C.; Titin, J.; Kitayama, K. Integration of carbon conservation into sustainable forest management using high resolution satellite imagery: A case study in Sabah, Malaysian Borneo. *Int. J. Appl. Earth Obs. Geoinf.* **2012**, *18*, 305–312. [CrossRef]
44. Langner, A.; Titin, J.; Kitayama, K. The Application of Satellite Remote Sensing for Classifying Forest Degradation and Deriving Above-Ground Biomass Estimates. In *Co-Benefits of Sustainable Forestry: Ecological Studies of a Certified Bornean Rain Forest*; Kitayama, K., Ed.; Springer: Tokyo, Japan, 2013; pp. 1–22. ISBN 9784431541417.
45. Poorter, L.; Bongers, F.; Aide, T.M.; Almeyda Zambrano, A.M.; Balvanera, P.; Becknell, J.M.; Boukili, V.; Brancalion, P.H.S.; Broadbent, E.N.; Chazdon, R.L.; et al. Biomass resilience of Neotropical secondary forests. *Nature* **2016**, *530*, 211–214. [CrossRef] [PubMed]
46. Chua, S.C.; Ramage, B.S.; Potts, M.D. Soil degradation and feedback processes affect long-term recovery of tropical secondary forests. *J. Veg. Sci.* **2016**, *27*, 800–811. [CrossRef]

Article

Spatial Pattern and Factor Analyses for Forest Sustainable Development Goals within South Korea's Civilian Control Zone

Jinwoo Park [1] and Jungsoo Lee [2,*]

1 Inter-Korean Forest Research Team, Division of Global Forestry, National Institute of Forest Science, Seoul 02455, Korea; source0310@korea.kr
2 Department of Forest Management, Division of Forest Sciences, College of Forest and Environmental Sciences, Kangwon National University, Chuncheon 24341, Korea
* Correspondence: jslee72@kangwon.ac.kr; Tel.: +82-33-250-8334

Received: 27 July 2018; Accepted: 26 September 2018; Published: 29 September 2018

Abstract: The United Nations' Sustainable Development Goals (SDGs) offer specific guidelines for improving sustainable forest management, especially Goal 15. Goal 15 protects, restores and promotes the sustainable use of land ecosystems, manages forests sustainably, prevents was against desertification, stops and reverses land degradation and prevents biodiversity loss. The Civilian Control Zone (CCZ) south of the Demilitarized Zone (DMZ) separating North and South Korea has functioned as a unique biological preserve due to traditional restrictions on human use but is now increasingly threatened by deforestation and development. We used hot spot analysis and structural equation modeling (SEM) to analyze spatial patterns of forest land use and land cover (LULC) change and variables influencing these changes, within the CCZ. Remote sensing imagery was used to develop land cover classification maps (2010 and 2016) and a GIS database was established for three change factors (topography, accessibility and socioeconomic characteristics). As a result of Hotspot analysis, Hotspots of change were distributed mainly due to agricultural activities and the development of forest and expansion of villages. Subsequent factor analysis revealed that accessibility had greater influence (−0.635) than the other factors. Among the direct factors, change to bare land had the greatest impact (−0.574) on forest change. These results shed light on forest change patterns and causes in the CCZ and provide practical data for efficient forest management in this area with regards to the SDGs.

Keywords: forest land change; land change patterns; Civilian Control Zone; DMZ; sustainable development goals (SDGs); forest management; structural equation modeling (SEM); factor analysis

1. Introduction

In September 2000, the United Nations (UN) presented Millennium Development Goals (MDGs) aimed at enhancing the quality of human life [1]. However, these MDGs set no concrete objectives for solving fundamental problems posed by factors hampering efforts to enhance quality of life. The 2012 UN Conference on Sustainable Development (UNCSD), "Rio + 20," produced the outcome document "The Future We Want" which set a new sustainable development agenda called Sustainable Development Goals (SDGs) [2]. In 2015, the UN adopted the "2030 Agenda for Sustainable Development" containing 17 SDGs and 169 targets [3]. The SDGs' primary goal is eradicating poverty, with a focus on balanced economic, social and environmental developments.

A broad range of forest-related issues, from forest management to forest-added value, were considered in multiple SDGs including Goal 3 (Good Health and Well-being), Goal 6 (Clean Water and Sanitation), Goal 13 (Climate Action) and Goal 15 (Life on Land) [4]. Goal 15, "Protect, restore and promote sustainable use of terrestrial ecosystems, sustainably manage forests, combat desertification and halt and reverse land degradation and halt biodiversity loss," is particularly associated with forestry; its targets emphasize the value of forests for biodiversity conservation along with marine and coastal areas [5].

In the Korean Peninsula, the area along the demilitarized zone between North and South Korea (DMZ) is valued as a treasure house of biodiversity together with the Baekdudaegan (Baekdu Mountains, the longest mountain chain running along the Korean Peninsula) and coastal islands. The DMZ was designated as one of the axes of the "Ecological Network of the Korean Peninsula" by the Ministry of Environment in 2002 [6]. The area is divided into the DMZ itself (about 4 km wide) and the Civilian Control Zone (CCZ), an area 5–20 km wide extending from the DMZ's Southern Limit Line to the Civilian Control Line further south. In the past, civilians were not allowed to enter the Civilian Control Zone (except during guarded tourist expeditions) and so its intact natural environment and ecosystems have been highly valued together with the DMZ [7]. However, as active government-led development programs are being implemented in the CCZ, its forests are exposed to increasing deforestation and forest degradation [8]. These processes degrade plant and animal habitats and disturb the ecological and environmental functions of the area, counteracting efforts to achieve the SDGs [9]. With increasing socioeconomic activities in the CCZ, the ongoing change from forest to non-forest land use is becoming increasingly complex [10]. The forests in the CCZ encompass a broad swath across the Korean Peninsula and proper understanding of forest land use and land cover (LULC) change factors is essential for integrated planning and management [11–14].

LULC change does not take place independently of the surrounding environment; it is therefore important to consider associations and correlations with adjacent areas when analyzing its causes [15]. Jeong et al. [16] emphasized the importance of understanding the cluster and diffusion aspects of LULC distribution and identifying the underlying social, economic and environmental factors when setting up land-management plans. Of the various analytical approaches used to explore this topic, Hot spot analysis has been increasingly prioritized as an intuitive spatial-statistical analysis method compared to more complicated calculation methods [17]. This approach, developed by Ord and Getis [18], expresses the similarities and differences of spatial values measured and summarized for clustered locations [19]. For example, Lee [20] used remote sensing data to identify the relationship between LULC change and a vegetation index. Choi et al. [21] found that Hot spot analysis was useful for exploring the clustered factors related to LULC change and socioeconomic data but also noted the method's limitations with regards to determining the effects of individual factors. Analyzed various influential factors for LULC change and explored the differing effects of these factors [22].

On the other hand, structural equation modeling (SEM) is widely used in social humanities studies for its ability to compare correlations between various factors and to determine the inter-factor differences in impact [23–25]. For example, Jang and Kim [26] evaluated the impact of development density for various land types on the land's value and Asadi et al. [27] evaluated the impact of economic, social and environmental factors on cropland changes. To date, however, most related studies have used cross-sectional data rather than time-series data and hardly any studies have performed comprehensive spatial data analysis.

In this study, we have conducted studies forest management and various factors influencing for DMZ forest, the only area in the world. We used Hot spot analysis to analyze patterns of forest LULC change and SEM to determine the effects of various factors influencing forest LULC change. Our results provide practical data for setting up comprehensive forest management programs in the CCZ as part of ongoing efforts to achieve SDGs.

2. Materials and Methods

2.1. Study Site

The CCZ was designated by Article 2(1) of the "Special Act on Management of Mountainous Districts North of The Civilian Control Line" and Article 5(2) of the "Protection of Military Bases and Installations Act" [28]. It extends over an area of ~120,000 ha, accounting for ~18% of the administrative divisions intersecting the DMZ area (Figure 1). The CCZ has a broad distribution of land use for military facilities due to its intrinsic nature but recent years have seen steady LULC changes related to the development of tourist attractions [29]. The Gyeonggi-do part of the CCZ is flat and low (<300 m above sea level) with intensive agricultural activities taking place on a residential or commuting basis. The Gangwon-do part of the CCZ is predominantly mountainous (≥1000 m above sea level) except for Cheorwon-gun [30].

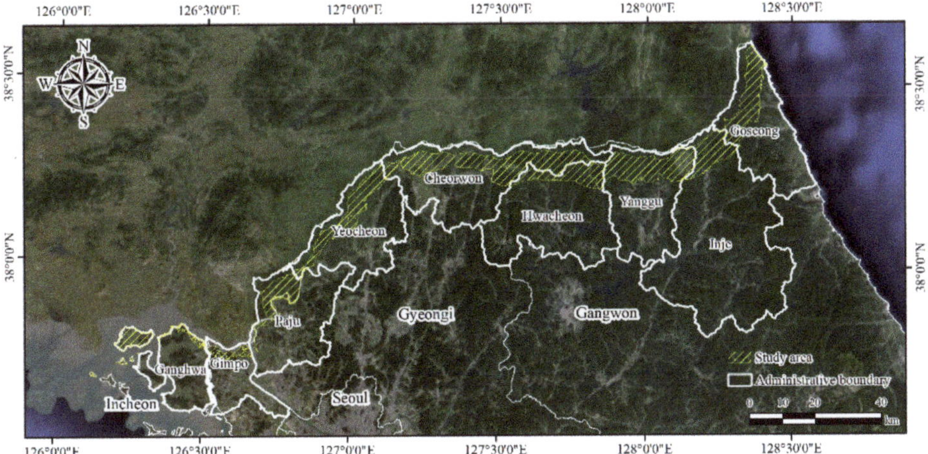

Figure 1. Study Site: The Civilian Control Zone (CCZ)) south of the demilitarized zone separating North and South Korea.

2.2. Materials

We used three types of data for our analyses: remote sensing (RS), geographic and statistical (Table 1). The RS data consisted of 2010 and 2016 land cover maps produced by applying maximum-likelihood supervised classification to medium-resolution satellite imagery using Envi 5.0 software. The land cover map was made using three scenes (path 116/row 33, path 116/row 34, path 115/row 33) and using Landsat 5 and Landsat 8 images. The classification accuracy of the land cover map was as follows, Overall accuracy: 93, Kappa: 0.91 in 2010, Overall accuracy: 943, Kappa: 0.92. The classification accuracy of the land cover maps was high in both overall accuracy and Kappa. Reference data for the accuracy of classification maps were used for field survey and Google earth images. The geographic data consisted of geographic information system (GIS) layers for administrative boundaries map provided by the Korea Forestry Service, the Southern Limit Line provided by the Gangwon Development Research Institute and CCZ, Digital topographic maps (scale of 1:25,000), cadastral valuation data and local population data provided by the Ministry of Land, Infrastructure and Transport. The statistical data consisted of tourist attraction information provided by the Korea Tourism Organization as spatial information. Typical tourist attractions of CCZ are "Unification Observatory", "DMZ Museum" and "Dam of Peace".

Table 1. Data used for analyzing the forest LULC patterns and change factors.

Type	Data	Remark
Remote sensing	Land cover maps - 2010: Landsat (2010.06.04., 2010.05.25.) - 2016: Landsat (2016.05.19., 2016.05.28.)	Produced using supervised classification with satellite imagery
Geographic	Administrative boundaries map	Korea Forestry Service
	Southern Limit Line	Gangwon Dev. Research Institute
	Civilian Control Zone	Ministry of Land, Infrastructure and Transport
	Digital topographic maps	
	Mean cadastral valuation (as of 2016 year)	
	Local population data (as of 2013 year)	
Statistical	Tourist attraction	Korea Tourism Organization

2.3. Methods

We analyzed the spatial patterns associated with forest LULC change and related factors in an effort to seek efficient ways to achieve the SDGs for forests within the CCZ. Areas that underwent change from forested to non-forest landed (forest change areas) were extracted by superimposing the 2010 and 2016 land-cover maps. A GIS database was then established by spatial assessment along with related change factors such as topography (slope and elevation), accessibility (access to roads, buildings and the CCZ) and socioeconomic characteristics (cadastral value, local population and tourist attractions). This was then used for the forest change pattern and change factor analyses using Hot spot analysis and SEM, respectively (Figure 2). Both assessments used the basic geographic unit ri (village), "ri" is the smallest administrative division in South Korea.

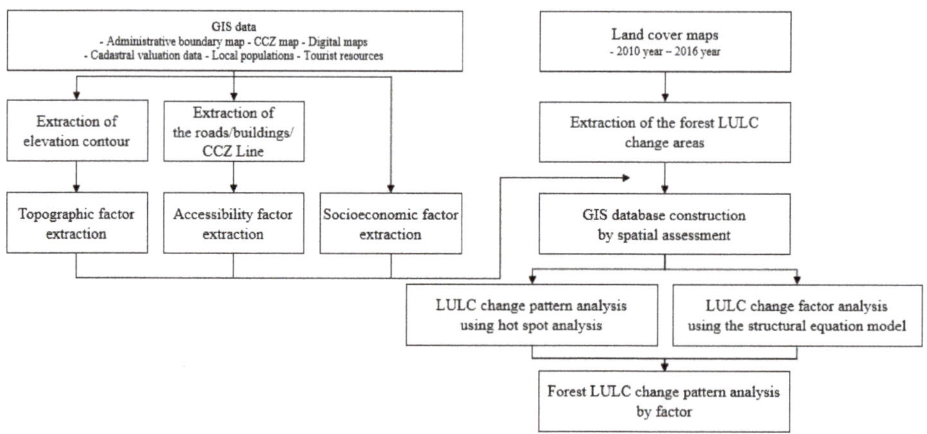

Figure 2. Flow chart of the forest LULC pattern and change factor analyses.

2.3.1. Forest LULC Change Spatial Analysis Using Spatial Statistics

In this study, spatial autocorrelation analysis (Hotspot analysis) was performed to analyze the pattern of forest land changing to other land.

LULC change is not independent of the surrounding environment, being influenced by adjacent land [20,31]. Therefore, we quantified interactions with surrounding areas using spatial autocorrelation, which measures the spatial similarity and/or dissimilarity of different regions in terms of proximity [32].

We used the Getis-Ord *Gi**, developed by [18], as the spatial autocorrelation index. This statistic calculates the proportions of the reference and neighboring spaces' values and measures the degree of spatial association between physical objects based on the concentration of weights. Getis-Ord *Gi** expresses spatial clusters of high and low values as hot and cold spots, respectively. When distribution patterns are randomly dispersed (without clusters) the values are close to zero [33].

$$Gi^* = \frac{\sum_{j=1}^{n} w_{ij}x_i - \overline{X}\sum_{j=1}^{n} w_{ij}}{s\sqrt{\frac{n\sum_{j=1}^{n} w_{ij}{}^2 - \left(\sum_{j=1}^{n} w_{ij}\right)^2}{n-1}}}$$

x_i = attribute value of i

w_{ij} = spatial weight (spatial weight matrix value)

s = standard deviation

n = total number of cases

i and i are adjacent = 1

i and i are not adjacent = 0

2.3.2. Forest LULC Change Factor Analysis Using SEM

SEM Background and Selection of Variables

SEM combines confirmatory factor analysis and path analysis, stemming from psychometrics and econometrics, respectively, through which causation and correlation between variables can be tested [34]. SEM is useful for estimating complicated causal relationships among multiple variables and simultaneously explaining direct and indirect effects between observable and unobservable ("latent") variables [35,36]. Furthermore, inter-variable relationships are displayed in schematic diagrams, allowing intuitive understanding of the analysis results [37]. In this study, SEM-based factor analysis involved multiple steps: selecting measurement variables with the potential to influence forest LULC, performing exploratory factor analysis to assess the inter-variable effects, building a study model based on the relationships among the measurement variables and analyzing the factors for forest LULC change by examining the causal relationships and associations among the variables.

SEMs have a structural component (path analysis) and a measurement component (factor analysis), in which the causal relationships are shown between dependent (endogenous) and independent (exogenous) variables and latent (constructs or factors) and observed (measured) variables, respectively. In the model we built for this study, the extent of forest LULC change per unit area was used as the measurement (dependent) variable. Latent variables including topographic (elevation (m) and slope (°)), accessibility (m) (to roads, buildings and the CCZ) and socioeconomic (population distribution, tourist attractions and mean cadastral value (₩)) factors were selected on the assumption that they could influence LULC change [12,38,39]. Elevation and slope information are constructed using digital topographic maps and accessibility is the distance (Unit: m) from the forest change areas of roads, buildings and CCZ. Population distribution and tourist attractions units are numbers and unit of mean cadastral value is won (₩). The input data for these observable variables were calculated as averages by ri, except for tourist attractions, which used the total number. The units of the input variables were homogenized and the final variables selected by means of exploratory factor analysis [40].

SEM Construction and Assessment of Model Fit

Figure 3 presents the SEM for the forest LULC change area. The two left-hand columns show the dependent variables consisting of five measurement variables and one latent variable. The two right-hand columns show the independent variables consisting of three latent variables and eight measurement variables.

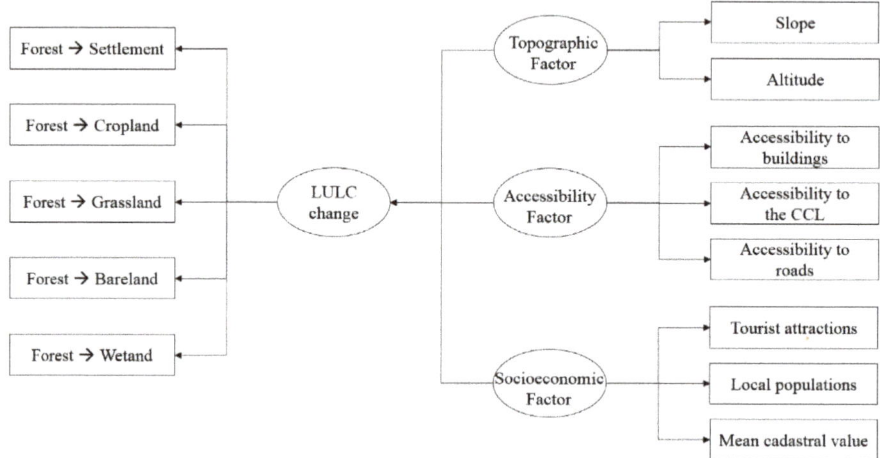

Figure 3. Framework of the SEM used in this study.

The SEM was evaluated using absolute and incremental fit indices (Table 2) to test how well the observed data fit into the model and how closely the measurement and latent variables were related to one another. An absolute fit index tests how well the proposed model predicts the sample covariance matrix and an incremental fit index measures the improvement in fit of the proposed model in comparison with the basic model [34,41].

Table 2. Fit indices and cutoff criteria used for SEM evaluation.

Model Fit Index		Cutoff Point
Absolute Fit	X^2 (CMIN)	$p < 0.05$
	Normed X^2 (CMIN/DF)	≤ 3
	Goodness of Fit Index (GFI)	≥ 0.9
	Standardized RMR (SRMR)	≤ 0.08
Incremental Fit	Normed Fit Index (NFI)	≥ 0.9
	Comparative Fit Index (CFI)	≥ 0.9

3. Results and Discussion

3.1. Spatial Pattern Analysis of the Forest Change Area

Statistically significant Hot spots for land-use change from forest to settlement in Ganghwa-gun, Gimpo-si, Yeoncheon-gun and Cheorwon-gun (Figure 4a). Jogang-ri in Gimpo-si and Naka-ri in Yeoncheon-gun displayed especially high Hot spot densities with Z-scores of 3.40 and 3.02, respectively. Frequent intense clusters of cold spots in Paju-si, Yeoncheon-gun and Cheorwon-gun also indicated the high rate of change from forest to settlement in these areas. Sung et al. [29] noted that deforestation in the CCZ is primarily due to the development of cropland and villages in Gyeonggi-do and the western part of Gangwon-do. Similarly, Lee [42] reported that the distribution density of villages in cities and counties with Hot spots accounted for about 74% of the overall distribution of villages in the CCZ. Village construction and expansion is especially intensive in Paju-si, Yeoncheon-gun and Cheorwon-gun, where agricultural activities are allowed on a residential or commuting basis.

Statistically significant Hot spots for land-use change from forest to cropland t in Gangnae-ri in Yeoncheon-gun, Woechon-ri in Cheorwon-gun and Mandae-ri in Yanggu-gun (Figure 4b). The latter displayed an especially high Hot spot density with a Z-score of 4.24. Hae'an-myeon and the surrounding area are characterized by intensive farming, with cropland accounting for 37.2% of the total area. Agricultural activities in this area are also diversifying into flowers, decorative plants, fruits and so forth, which continues to drive the conversion of forest into fields and orchards [43]. Park and Nam [44] reported on the continuously increasing land used for ginseng cultivation in this area since it hosted the "Gaeseong Ginseng Festival" in 2005. Sung et al. [29] also noted that expanding rice paddies and fields are a major cause of deforestation and emphasized that ginseng cultivation fields not only damage the primarily affected terrain but also cause secondary damage to surrounding terrain, weakening soils and degrading other vegetation. The increasing damage to forest edges in affected areas of Ganghwa-gun, Paju-si and Yeoncheon-gun can thus be attributed to the active ginseng cultivation operations in these areas in conjunction with broadly scattered small-scale agricultural land intermingled with forest.

Statistically significant Hot spots for land-use change from forest to grassland in Ganghwa-gun, Paju-si and Yeoncheon-gun (Figure 4c). The highest Hot spot densities by Z-score occurred in Cho-ri and Banjeong-ri in Paju-si and Majeon-ri and Dapgok-ri in Yeoncheon-gun (Figure 4c).

Statistically significant Hot spots for land-use change from forest to bare land in Ganghwa-gun, Yeoncheon-gun and Cheorwon-gun (Figure 4d). High-density Hot spots appeared in Gogu-ri in Ganghwa-gun and Hwoengsan-ri in Yeoncheon-gun (Z-scores: 3.94 and 3.59, respectively). Song et al. [45] reported that the LULC change rate was rapidly increasing after the government announced its "Master Plan for Border Area Land Use Development" in 2011. "Master Plan for Border Area Land Use Development" is a plan announced by the Ministry of the Interior in 2011. This plan sets out the vision and goals for the development of CCZ. However, because this plan prioritizes development, forest researchers are worried that this plan will increase deforestation and degradation. Sung et al. [29] reported on continuous illegal clearing of trees and vegetation for soybean and ginseng cultivation in Yeoncheon-gun and Cheorwon-gun. The change from forest to bare land within the CCZ from Ganghwa-gun to Cheorwon-gun is thus ascribable to such land-use projects and illegal cropland development.

No Hot spots appeared in the spatial pattern indicating change from forest to wetland but cold spots appeared in Gangnae-ri in Yeoncheon-gun, Mandae-rin In Yanggu-gun and Seohwa-ri in Inje-gun (Figure 4e). Small-scale increases in wetland were at first assumed to be the result of precipitation but rainfall data for June 2010 and May 2016, the periods corresponding to the RS data used for the analysis, showed insufficient rainfall to cause significant changes in land cover during these periods [46,47]. A conversion of forest to wetland was interpreted as artefact, because no factual evidence was found in the field.

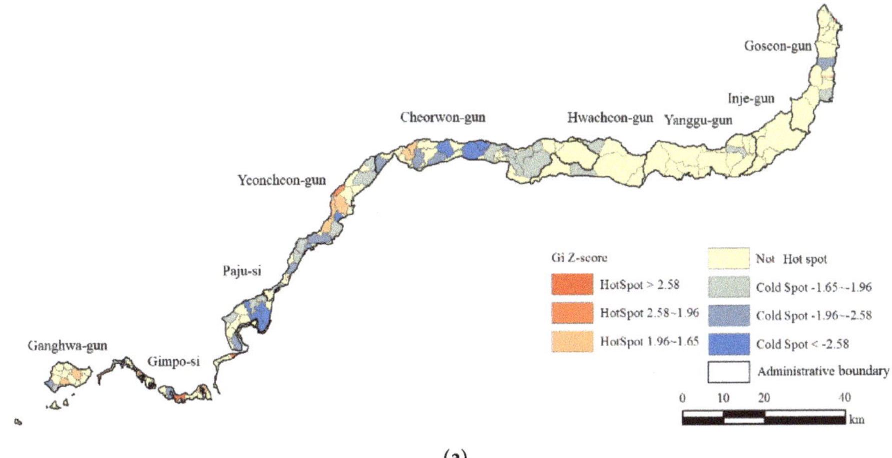

(a)

Figure 4. *Cont.*

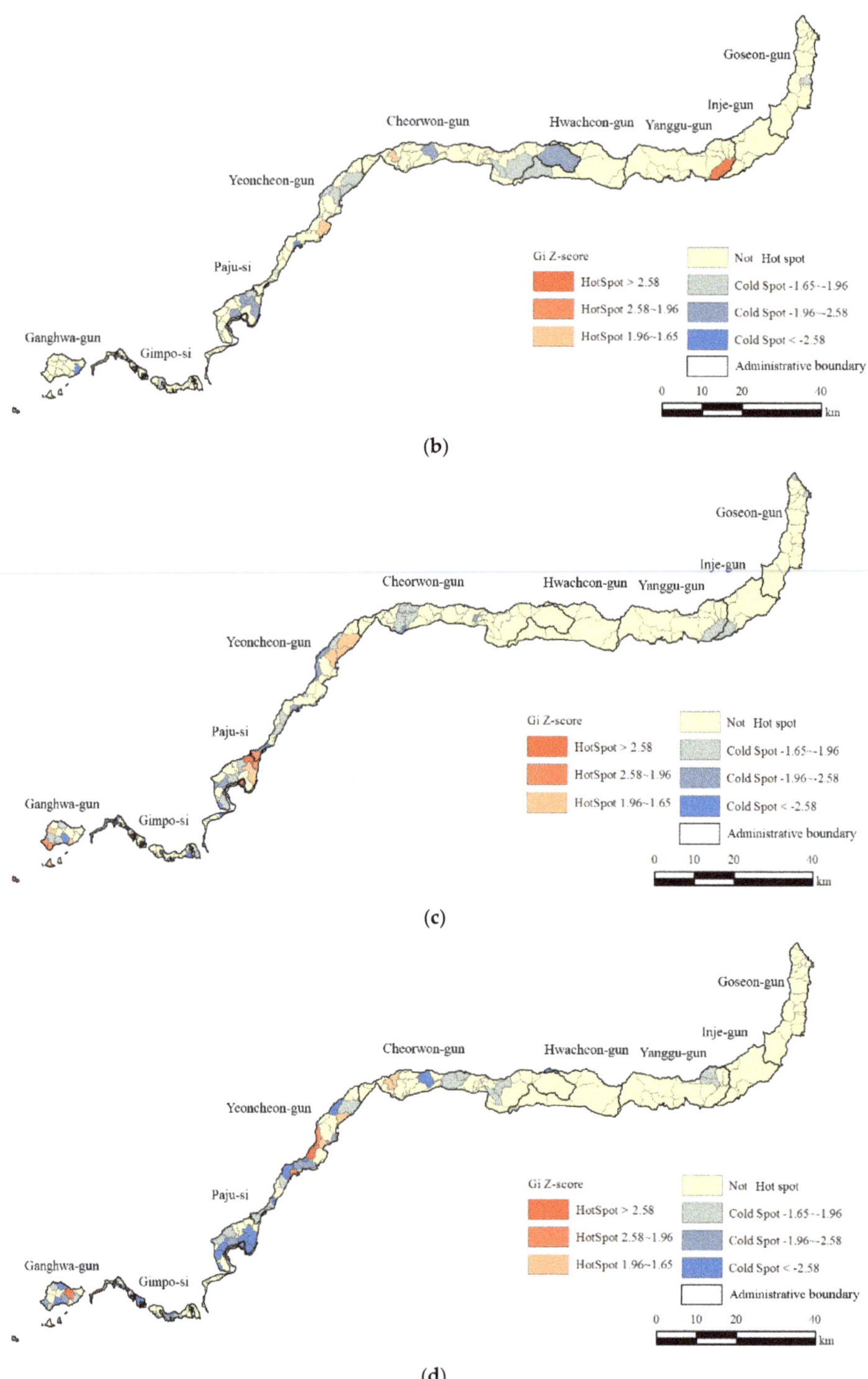

Figure 4. *Cont.*

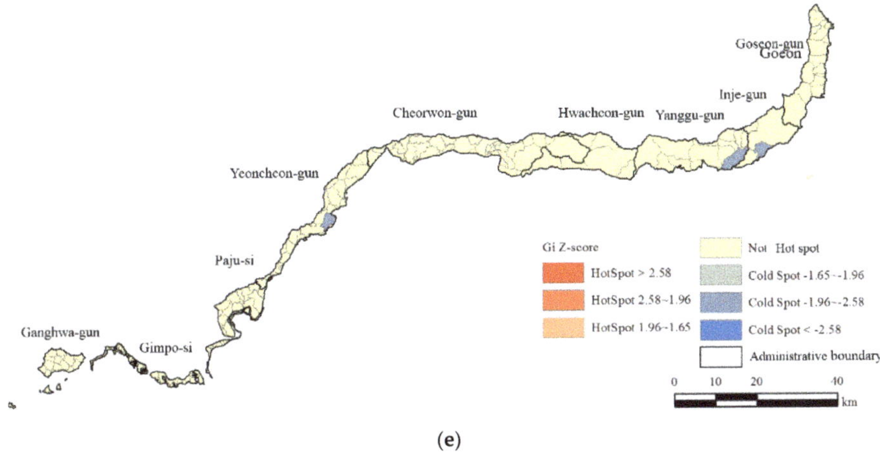

(e)

Figure 4. Hot spot distribution maps of forest LULC change: (**a**) forest to settlement; (**b**) forest to cropland; (**c**) forest to grassland; (**d**) forest to bare land; (**e**) forest to wetland.

3.2. Determination of Variables by Factor Analysis and SEM Fit Test

Factor analysis is a statistical method used to test the validity of variables to be measured, in which variables with low validity are excluded from analysis or modified [48]. We used the Kaiser-Meyer-Olkin (KMO) test to assess correlations among variables for factor analysis and Bartlett's Test to determine the suitability of the factor analysis model. In general, a KMO value of 0.9 or above was deemed excellent, 0.8–0.9 good, 0.7–0.8 reasonable and 0.5 or less unacceptable [48]. The KMO and Bartlett's Test values for this study's variables were found to be good, at 0.815 and 0.000, respectively, demonstrating the suitability of the model used.

Commonality is the rate at which a specific variable is explained for various factor. Variables with a commonality of less than 0.4 are excluded from the factor analysis [49]. The variables used in this study were found to be suitable for factor analysis with communalities in excess of 0.4 (Table 3). All model fit indices were within acceptable ranges, so the model was judged to be suitable for hypothesis testing and quantifying causal relationships [37] (Table 4).

Table 3. Similarities of the variables extracted by factor analysis.

Variables Used for Factor Analysis		Communality
Forest Area Change	Forest to settlement	0.535
	Forest to cropland	0.654
	Forest to grassland	0.569
	Forest to bare land	0.446
	Forest to wetland	0.752
Topographic Factor	Slope	0.790
	Elevation	0.800
Accessibility Factor	Accessibility to buildings	0.769
	Accessibility to the CCZ	0.731
	Accessibility to roads	0.752
Socioeconomic Factor	Tourist attractions	0.652
	Local populations	0.630
	Mean cadastral value	0.716

Table 4. Fit indices and cut-off criteria of the study model.

Model Fit Index	Analysis Result	Cut-Off Point	Fit/Unfit
CMIN/p-value	72.303/0.040	$p < 0.05$	Fit
CMIN/DF	1.364	$\leq$3	Fit
GFI	0.952	$\geq$0.9	Fit
SRMR	0.061	$\leq$0.08	Fit
NFI	0.905	$\geq$0.9	Fit
CFI	0.972	$\geq$0.9	Fit

3.3. Factor Analysis of Forest LULC Change

The direct effect of the topographic factor on forest area change was −0.503, implying that topography had an inversely proportional effect on the extent of forest change. The greatest indirect effect was shown in the change to grassland (−0.428). Elevation was found to have a stronger impact (0.935) than slope (Figure 5a). Unlike grassland and bare land, the indirect effects of topography on changes to settlement, cropland and water were found to be minimal. Sung et al. [29] reported that increasing agricultural activity such as Ginseng cultivation is the main cause of vegetation deterioration and increasing bare land in the CCZ. Ganghwa-gun, Paju-si and Yeoncheon-gun were reported to be particularly affected by intensive agricultural activities. This is consistent with the factor analysis result that changes to grassland and bare land were more intense in the low-slope and low- elevation regions of the western CCZ.

The direct effect of the accessibility factor was −0.635, higher than the other two factors (topographic and socioeconomic) (Figure 5b). Accessibility had a negative effect on forest change, implying that change from forest cover to other land use increases in proportion with proximity to roads, buildings and the CCZ. The greatest indirect effect was shown in the change to bare land (−0.574), followed by grassland and cropland. The accessibility factor showed similar characteristics to the topographic factor, with higher impact on the change to grassland and lower impact on the change to bare land. This is consistent with the findings of Sung et al. [29] that deforestation and degradation increases in proportion to the level of forest land used for military facilities, strategic roads and villages.

The direct effect of the socioeconomic factor was 0.053, with forest change increasing with increases in tourist attraction distribution, local population and cadastral value (Figure 5c). The greatest indirect effect was shown in the change to grassland but the socioeconomic factor had very low effect on forest change overall, below 0.100 in both direct and indirect effects, presumably due to the intrinsically low numbers of tourist attractions and population in the CCZ.

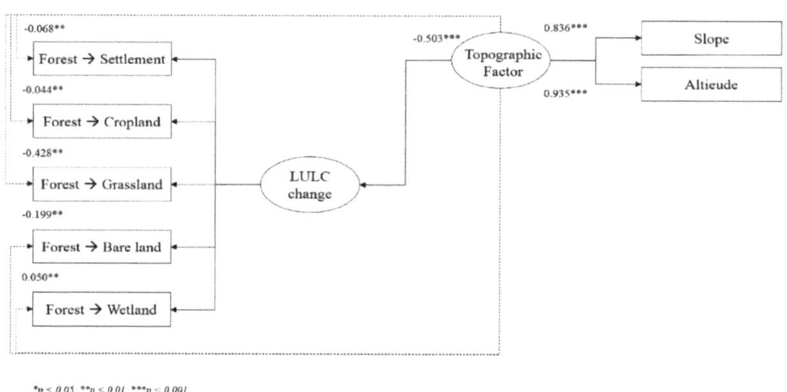

(a)

Figure 5. *Cont.*

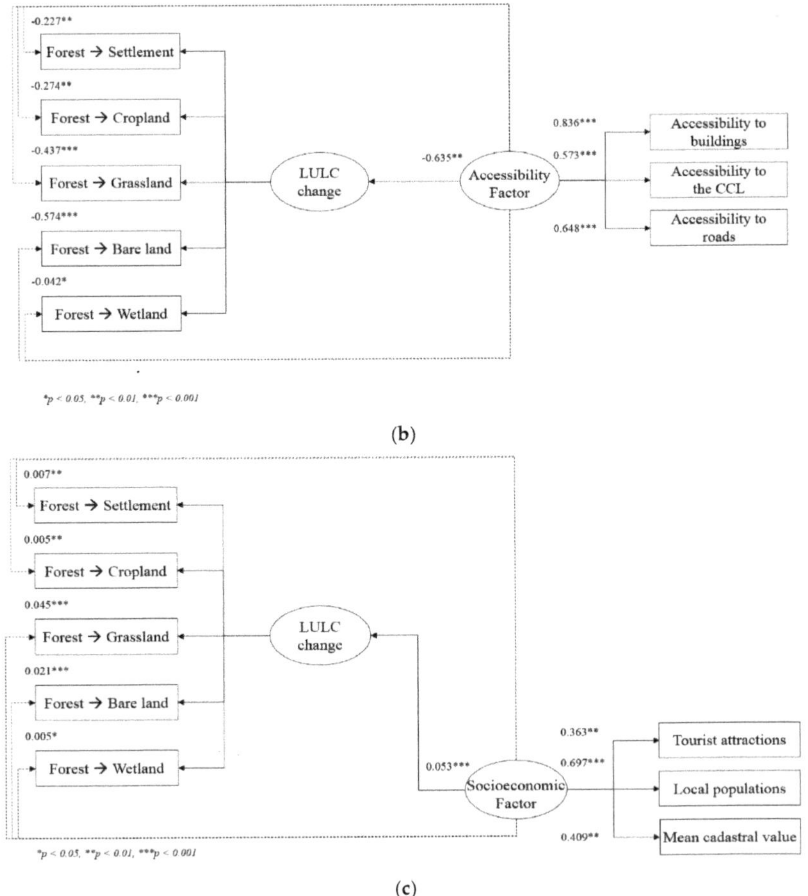

(b)

(c)

Figure 5. Factor analysis of the forest change area using the structural model: (**a**) topographic factor; (**b**) accessibility factor; (**c**) socioeconomic factor.

4. Conclusions

To achieve the UN's SDG Goal 15 (forest management) within South Korea's CCZ, it is essential to detect areas requiring restoration and management. Classification of remote sensing imagery showed that the predominant spatial patterns in this area's forest LULC change were from forest to settlement, cropland, grassland and bare land.

Our Hot spot analysis enabled an intuitive analysis of these patterns and facilitated the identification of areas requiring intensive management. High distribution densities of such changes occurred in Ganghwa-gun, Gimpo-si, Paju-si and Yeoncheon-gun. The main causes of forest area change were agricultural activities (rice paddies and fields) and the development and expansion of villages. Increasing degradation at the forest edge caused by increasing proportions of cropland, grassland and bare land was directly associated with increasing deterioration of vegetation and deforestation. Due to the unique nature of the CCZ, deforestation and degradation, once done, is difficult to reverse.

Structural equation modeling determined the effects of factors associated with forest change. Of the three factors extracted (topography, accessibility and socioeconomic characteristics), the accessibility factor (proximity to buildings, roads and the CCZ) showed the strongest effect on forest change, above all on changes to bare land. The topographic factor (elevation and slope) also had great influence on forest change, above all on changes to grassland. These factors are therefore strong indicators of high degrees of deforestation and degradation.

According to previous studies, deforestation and degradation in the CCZ increases in proportion to the level of forest changes for the purposes of building military facilities, strategic roads, villages and ginseng cultivation, resulting in accelerated damage to the ecosystems in the CCZ. Such damaged areas are also problematic because they can contribute to natural disasters such as floods due to the degraded and deforested landscape. Achieving the SDGs' forest management goals is important for both forest conservation and providing benefits such as biodiversity, carbon sinks, water purification and storage and recreation. These results provide practical data for preparing direct and indirect measures necessary for improved forest management programs in the CCZ.

Author Contributions: J.L. conceived and designed the experiments; J.P. performed the experiments and analyzed the data; J.P. and J.L. wrote the paper together. J.L. contributed to the modification. All authors have read and approved the final manuscript.

Funding: This study was carried out with the support of 'R&D Program for Forest Science Technology (Project No. "2017045B10-1819-BB01")' provided by Korea Forest Service (Korea Forestry Promotion Institute).

Conflicts of Interest: The authors declare no conflict of interest.

References

1. Park, S. A Study on the Millennium Development Goals and Sustainable Development Goals of UN for Enhancing the Quality of Human Life. *J. Korean Soc. Qual. Manag.* **2014**, *42*, 529–542. (In Korean) [CrossRef]
2. Pisano, U.; Lepuschitz, K.; Berger, G. Fostering Urban Sustainable Development-Potentials and Challenges for National and European Policy. In *10th ESDN Workshop-Background and Discussion Paper*; European Sustainable Development Network: Brussels, Belgium, 2014.
3. United Nations. Transforming Our World: The 2030 Agenda for Sustainable Development. A/RES/70/1. Available online: https://sustainabledevelopment.un.org/post2015/transformingourworld (accessed on 22 April 2018).
4. Forests and SDGs: Taking a Second Look. Available online: http://www.wri.org/blog/2017/09/forests-and-sdgs-taking-second-look (accessed on 24 April 2018).
5. Moon, J.; Kim, N.; Song, C.; Lee, S.G.; Kim, M.; Lim, C.H.; Cha, S.E.; Kim, G.; Lee, W.K.; Son, Y.; et al. Analysis on the Linkage between SDGs Framework and Forest Policy in Korea. *J. Clim. Chang. Res.* **2017**, *8*, 425–442. (In Korean) [CrossRef]
6. Jeon, S.; Lee, M.; Kang, B.; Shin, J. Study for Building Ecological Network in East-North Asia. *J. Environ. Policy* **2009**, *8*, 1–26. (In Korean)
7. Seo, C.; Jeon, S. Landcover Analysis of DMZ and the Vicinity Using Remote Sensing and GIS Techniques. *J. Environ. Impact Assess.* **1998**, *7*, 11–22. (In Korean)
8. Park, E.; Nam, M.; Park, M. *DMZ Eco Peace Villages: Basic Survey and Development Strategies*; Gyeonggi Research Institute: Suwon, Korea, 2012; pp. 4–15. (In Korean)
9. You, J.H.; Jung, S.G.; Oh, J.H. Construction of System on Assessment Indicators for Conservation of Sustainable Natural Ecosystem. *Korean J. Environ. Ecol.* **2005**, *19*, 287–298. (In Korean)
10. Kim, C.; Park, J.; Jang, D. A prediction of the Land-cover Change Using Multi-temporal Satellite Imagery and Land Statistical Data: Case Study for Cheonan City and Asan City, Korea. *J. Korean Geomorphol. Assoc.* **2011**, *18*, 41–56. (In Korean)
11. In, C. A Study on the Systematic Management Measures of Natural Resources in the Demilitarizes Zone: Based on the Demilitarizes Zone in Gyeonggi-Do. Master's Thesis, Dankook University, Yongin, Korea, August 2013.
12. Oh, Y.; Choi, J.; Bae, S.; Yoo, S.; Lee, S. A Probability Mapping for Land Cover Change Prediction using CLUE Model. *J. Korean Soc. Rural Plan.* **2010**, *16*, 47–55. (In Korean)
13. Tomao, A.; Quatrini, V.; Corona, P.; Ferrara, A.; Lafortezza, R.; Salvati, L. Resilient landscapes in Mediterranean urban areas: Understanding factors influencing forest trends. *Environ. Res.* **2017**, *156*, 1–9. [CrossRef] [PubMed]

14. Zhao, C.; Jensen, J.; Zhan, B. A comparison of urban growth and their influencing factors of two border cities: Laredo in the US and Nuevo Laredo in Mexico. *Appl. Geogr.* **2017**, *79*, 223–234. [CrossRef]
15. Kim, S.; Lee, J. The Effect of Spatial Scale and Resolution in the Prediction of Future Land Use using CA-Markov Technique. *J. KAGIS* **2007**, *10*, 57–69. (In Korean)
16. Jeong, K.; Moon, T.; Jeong, J. Hotspot Analysis of Urban Crime Using Space-Time Scan Statistics. *J. KAGIS* **2010**, *13*, 14–28. Available online: http://www.riss.kr/link?id=A82392455 (accessed on 24 April 2018). (In Korean)
17. Jeong, J.; Jun, M. Spatial Concentrations of the Elderly and Its Characteristics in the Seoul Metropolitan Area. *J. Korean Reg. Sci. Assoc.* **2013**, *29*, 3–18. (In Korean)
18. Ord, J.; Getis, A. Local spatial autocorrelation statistics: distributional issues and an application. *Geogr. Anal.* **1995**, *27*, 286–306. [CrossRef]
19. Brown, K.A.; Parks, K.E.; Bethell, C.A.; Johnson, S.E.; Mulligan, M. Predicting plant diversity patterns in Madagascar: Understanding the effects of climate and land cover change in a biodiversity hotspot. *PLoS ONE* **2015**, *10*, e0122721. [CrossRef] [PubMed]
20. Lee, J.; Kim, S.; Han, K.; Lee, Y. Hotspot Detection for Land Cover Changes Using Spatial Statistical Methods. *Korean J. Remote Sens.* **2011**, *27*, 601–611. (In Korean) [CrossRef]
21. Choi, J.; Park, W.; Kim, M. Relationship between land cover change and socio-economic change, and spatial query for time series data in the Saemangeum area. *Geogr. J. Korea* **2014**, *48*, 363–373. (In Korean)
22. Lee, T. Development and Application of Land Use Forecasting Model for Conurbanized Regions. Ph.D. Thesis, Chungbuk University, Cheongju, Korea, February 2015. (In Korean)
23. Lee, J. Satisfaction Factors Analysis of Freight Drivers Using Structural Equation Model. Master's Thesis, Hanyang University, Seoul, Korea, August 2015. (In Korean)
24. Van Acker, V.; Witlox, F.; Van Wee, B. The effects of the land use system on travel behavior: a structural equation modeling approach. *Transp. Plan. Technol.* **2007**, *30*, 331–353. [CrossRef]
25. Zhou, T. Understanding online community user participation: A social influence perspective. *Internet Res.* **2011**, *21*, 67–81. [CrossRef]
26. Jang, M.; Kim, T. An Analysis of Causal Relationship between the Density of Land Use Types, and Land Value Using PLS-Structural Equation Model: A Case of 25 Gu-Districts in Seoul. *Korean Urban Manag. Assoc.* **2012**, *25*, 65–87. (In Korean)
27. Asadi, A.; Barati, A.A.; Kalantari, K. Analyzing and modeling the impacts of agricultural land conversion. *Bus. Manag. Rev.* **2014**, *4*, 176–184.
28. Special Act on Management of Mountainous Districts North of The Civilian Control Line, Ministry of Government Legislation. Available online: http://www.law.go.kr/lsInfoP.do?lsiSeq=136901#0000 (accessed on 29 April 2018).
29. Sung, H.; Kim, S.; Kang, D.; Seo, J.; Lee, S. Analysis on the Type of Damaged Land in DeMilitarized Zone (DMZ) Area and Restoration Direction. *J. Korea Soc. Environ. Restor. Technol.* **2016**, *19*, 185–193. (In Korean) [CrossRef]
30. Park, E.; Jeon, S.; Nam, M.; Hong, S. *Ecosystem Destroying Factors and the Mitigation of Impacts in the Civilian Control Zone*; Gyeonggi Research Institute: Suwon, Korea, 2011; pp. 1–169. (In Korean)
31. Na, H.; Park, J.; Choi, J.; Lee, J. A Study on Spatial Characteristic of Damaged Forest Areas by Image Interpretation-Focused on DMZ region in Hwacheon and Cheorwon. *J. Assoc. Korean Photo-Geogr.* **2015**, *25*, 63–73. (In Korean)
32. Anselin, L.; Bera, A.K. Spatial dependence in linear regression models with an introduction to spatial econometrics. In *Handbook of Applied Economic Statistics*; Ullah, A., Ciles, D., Eds.; CRC Press: Boca Raton, FL, USA, 1998; pp. 237–290.
33. Park, J. Delimiting Urban Regeneration Districts Using Spatial Analysis. Master's Thesis, Kyungpook University, Daegu, Korea, February 2012. (In Korean)
34. Bae, B. *Structural Equation Modeling with Amos 19: Principles and Practice*; Chungram: Seoul, Korea, 2011; pp. 254–321. (In Korean)
35. Lee, H.; Lim, J. *Structural Equation Modeling with AMOS 18.0/19.0*; Jyphyunjae: Seoul, Korea, 2011; pp. 157–231. (In Korean)

36. Kim, J. An Analysis of the Changes in the Cause-and-Effect Relationships between Socio-Economic Indicators and the Road Network of Seoul Using Structural Equation Model. *J. Korean Geogr. Soc.* **2009**, *44*, 797–812. (In Korean)
37. Lee, K.; Jeong, J.; Heo, B.; Byeon, S. A Study on the Effect Analysis for the Regeneration Project for the Zones Vulnerable to Natural Disaster using Structural Equation Model. *J. Korea Water Resour. Assoc.* **2013**, *46*, 843–855. (In Korean) [CrossRef]
38. Yu, M.; Lee, M.; Choi, B. A Study on the Sustainable Urban Development Indicator—Focused on the environment-indicators. *Korean J. Hydrosci.* **2008**, *43*, 225–236. (In Korean)
39. Choi, D.; Suh, Y. Geographically Weighted Regression on the Environmental-Ecological Factors of Human Longevity. *J. Korea Soc. Geospat. Inf. Sci.* **2012**, *20*, 57–63. (In Korean) [CrossRef]
40. Kim, J. Development and Confirmation of the Theory of Learning Flow Processes: A Sequential Mixed Method of Grounded Theory and Structural Equation Modeling. Ph.D. Thesis, Chonbuk University, Jeonju, Korea, February 2013. (In Korean)
41. Kim, J. The Effects of Elderly Patients' Dental Satisfaction on Revisit Intention with the Application of SEM. Ph.D. Thesis, Inha University, Incheon, Korea, February 2013. (In Korean)
42. Lee, T. A Study on the Change of the Spatial Distribution of Villages in Civilian Control Zone by using GIS. Master's Thesis, Kangwon University, Chuncheon, Korea, August 2012. (In Korean)
43. Lee, W. Analysis of Temporal Change in Soil Erosion Potential at Haean-Myeon Watershed Due to Climate Change. Master's Thesis, Kangwon University, Chuncheon, Korea, August 2014. (In Korean)
44. Park, E.; Nam, M. Changes in Land Cover and the Cultivation Area of Ginseng in the Civilian Control Zone —Paju City and Yeoncheon County. *Korean J. Environm. Ecol.* **2013**, *27*, 507–515. (In Korean)
45. Song, J.; Lim, C.; Kim, D.; Park, M.; Ham, B.; Kim, M.; Lee, C. Forest Land Use Patterns in the Northern Area of the Civilian Passage Restriction Line. *J. Korea For. Recreat. Soc. Issues* **2013**, *4*, 805–807. (In Korean)
46. Korea Meteorological Administration. Monthly Weather Report (2010.6.). Available online: http://www.kma.go.kr/repositary/sfc/pdf/sfc_mon_201006.pdf (accessed on 29 April 2018).
47. Korea Meteorological Administration. Monthly Weather Report (2016.5.). Available online: http://www.kma.go.kr/repositary/sfc/pdf/sfc_mon_201605.pdf (accessed on 29 April 2018).
48. Song, J. *SPSS/AMOS Statistical Analysis Method*; 21st Century: Paju, Korea, 2009; pp. 254–278. (In Korean)
49. Seong, T. *Easy Statistical Analysis*; Hakjisa: Seoul, Korea, 2014; pp. 311–364. (In Korean)

Article

A Practical Index to Estimate Mangrove Conservation Status: The Forests from La Paz Bay, Mexico as a Case Study

Giovanni Ávila-Flores [1], Judith Juárez-Mancilla [2], Gustavo Hinojosa-Arango [3,*], Plácido Cruz-Chávez [2], Juan Manuel López-Vivas [1] and Oscar Arizpe-Covarrubias [1]

1 Academic Department of Marine and Coastal Sciences, Autonomous University of Baja California Sur, Baja 23037 California Sur, Mexico; giovanniavila@hotmail.com (G.Á.-F.); jmlopez@uabcs.mx (J.M.L.-V.); oarizpe@uabcs.mx (O.A.-C.)

2 Academic Department of Economics, Autonomous University of Baja California Sur, 23037 Baja California Sur, Mexico; juarez@uabcs.mx (J.J.-M.); pcruz@uabcs.mx (P.C.-C.)

3 CIIDIR Oaxaca, National Polytechnic Institute, 95060 Santa Cruz, Mexico

* Correspondence: ghinojosaar@conacyt.mx

Received: 30 November 2019; Accepted: 19 January 2020; Published: 23 January 2020

Abstract: Mangrove cover has declined significantly in recent years in tropical and subtropical areas around the world. Under this scenario, it is necessary to elaborate and implement tools that allow us to make estimations on their conservation status and improve their protection and support decision-making. This study developed an index using qualitative and quantitative data. The criterions used in the index were: (1) Remnant Vegetation Index, (2) Delphi Method Survey, and (3) Rapid Assessment Questionnaire. In turn, the weights of the criterions were defined using the analytical hierarchy process (AHP). Once the values of each criterion were obtained, the index was applied to 17 mangrove communities located in La Paz Bay, Mexico. Finally, according to their score, they were classified based on the IUCN Red List of Ecosystems. The results show that five communities were ranked in the category Minor Concern, eight in Little Threatened, one in Vulnerable, one in Endangered, and two were classified as Deficiency of Data. These results are slightly different from other studies in the region and validate this index as a proper method. Therefore, it could be applied to other sites, especially in areas with little information and/or scarce monetary resources.

Keywords: AHP; Delphi Method; GIS; MCDA; Sustainable Development; Wetlands

1. Introduction

1.1. Theoretical Background

Mangroves are recognized worldwide due to various resources and ecosystem services (ES) they provide [1,2]. Some of these ES are provision services (food, timber, medicines, etc.), regulation and maintenance services (protection against hurricanes, erosion control, CO_2 capture, etc.), and cultural services (aesthetic, symbolic, religious, etc.) [3,4]. The presence of coastal wetlands such as mangroves helps contribute to human wellbeing [5,6]. However, mangroves are one of the most threatened coastal ecosystems due to changes in environmental factors and impacts induced by human activities [7]. They are particularly vulnerable to degradation as a result of deforestation, aquaculture, agriculture, tourism, urbanization, and pollution from different sources [8]. Other factors, such as sediment dynamics, exotic species, and alteration of the hydrodynamics, also result in severe mangrove deterioration and habitat loss [9].

According to FAO data, about 3.6 million hectares of mangroves were lost in the 1980–2005 period, approximately 20% of the global mangrove cover [10]. Also, it is estimated that mangrove forests worldwide are disappearing between one to two percent each year, at a higher rate than rainforests or coral reefs [11]. Likewise, it is recognized that only 6.9% of the world's mangroves are in the category of a protected area network [12]. In México, the National Commission for the Use and Conservation of Biodiversity (CONABIO) through the Mexican Mangrove Monitoring System (SMMM) began a full examination of the mangroves in this country. The SMMM observed that in 1970–1980, Mexico presented a cover of 856,405 hectares but by 2015 the number of hectares of mangroves decreased to 775,555 [13]. This loss of mangroves has been observed in urban areas and in zones with a tourist-oriented vision. A well-known example is the destruction of "Tajamar" mangrove, in Cancun city, in 2016 [14].

Another example is the case of mangroves located in La Paz city. The capital of Baja California Sur state, has caused a remarkable anthropic pressure to their mangroves due to the growing urbanization [15]. Also, it is expected that tourism will increase in this city. This is evidenced in the list "52 Places to Go in 2020" of The New York Times, where La Paz was the only Mexican city nominated [16]. In this context, different forums, such as the "I Workshop on Mangroves of the Baja California Peninsula" (2005) and the "I Workshop of RAMSAR Sites of Baja California Sur" (2009) have highlighted the need to know the state of mangroves in La Paz bay. Therefore, the goal of this work was to develop an integrative index to evaluate the conservation status of mangroves in this region.

1.2. Study Area

La Paz Bay is in the southwest of the Gulf of California, at the southern portion of the Baja California Peninsula, in Mexico (Figure 1). This bay has diverse biotic and abiotic characteristics that made it a remarkable site with diverse ecosystems, such as beaches, dunes, seagrass beds, rocky reefs, and mangroves [17]. Its coastal zone harbors 17 mangrove communities, 16 of them within the proximity of La Paz city, and one more at the Espíritu Santo Archipelago. These mangroves are within protected areas such as the Espíritu Santo Archipelago National Park, the Balandra Flora and Fauna Protection Area, and the Islands of the Gulf of California Flora and Fauna Protection Area. They are also under the category of international protection, such as the Ramsar Sites "Humedales Mogote-Ensenada de La Paz" (Ramsar site no. 1816) and "Balandra" (Ramsar site no. 1767).

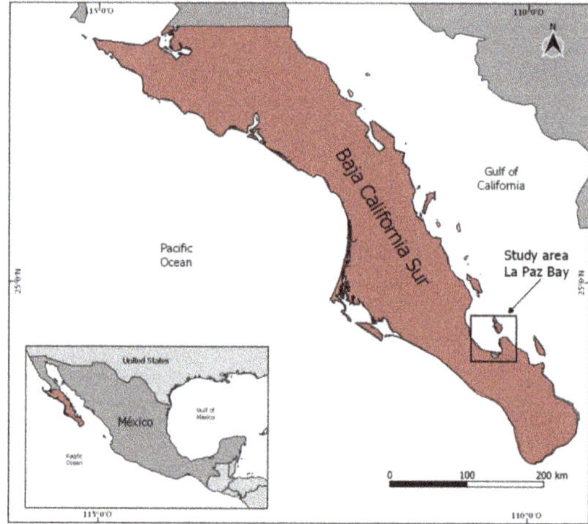

Figure 1. Location of La Paz Bay in the southern end of the Peninsula of Baja California, Mexico.

The 14 communities belonging to the Ramsar site No. 1816 are: Centenario-Chametla, Comitán, El Conchalito, El Mogote, Enfermería, Eréndira, Estero Bahía Falsa, Estero El Gato, La Paz-Aeropuerto, Palmira, Playa Pichilingue-Brujas, Unidad Pichilingue UABCS, Salinas de Pichilingue, and Zacatecas.

Two of these remaining communities belonging to Ramsar site no. 1767 are Balandra and El Merito. Finally, the mangrove Espíritu Santo Archipelago is located within another protected area, it is in the national park category and its name is homonymous to that community (Figure 2).

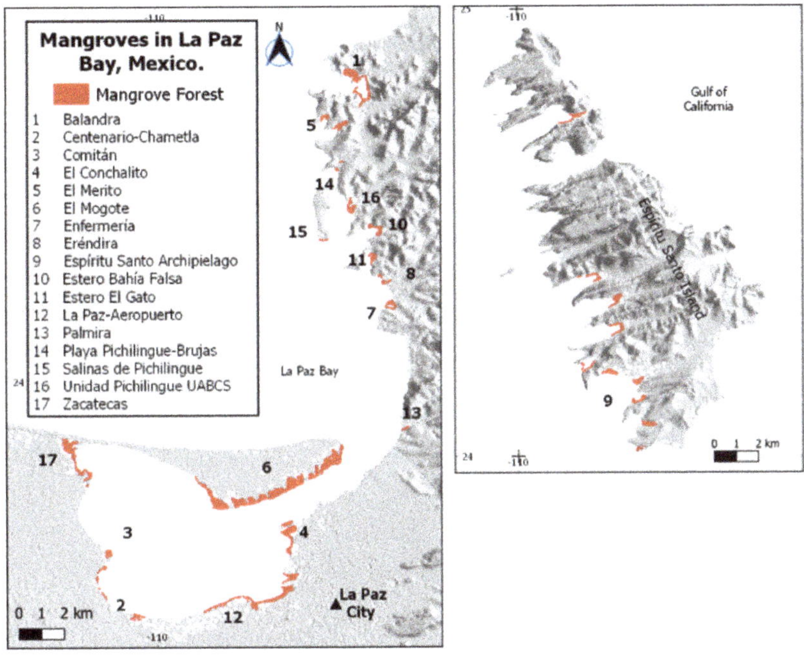

Figure 2. Location of mangroves placed in La Paz Bay, Mexico.

2. Materials and Methods

2.1. Study Design

The development and implementation of the Mangrove Conservation Status Index (MCSI) consisted of five phases carried out between January and September 2019. First, the remnant vegetation index (RVI), Delphi method survey (DMS), and the rapid assessment questionnaire (RAQ) were conducted at the 17 mangrove areas during January–February 2019, followed by the use of the analytical hierarchy process (AHP) to estimate the weights of each one of these indicators. We conducted a spatial analysis on open-source GIS software QGIS (version 3.4.4) to calculate the RVI, designed a questionnaire for the RAQ, and applied surveys to local mangrove experts for the DMS. Finally, we estimated the Mangrove Conservation Status Index for all the mangroves.

2.2. Index Development and AHP

For the construction of the MCSI, three components were selected, as these are widely used in various environmental analyses (RVI, DMS, and RAQ). For example, several studies have carried out comparisons between mangrove cover between different years, globally and nationally and locally. Also, the application of the Delphi method in environmental modeling is considered as a useful tool, with various studies focused on mangroves. Finally, the rapid assessment tool has also been used in forest and wetland analysis. Once the index components were selected, the following formula was generated:

$$\mathbf{MCSI} = (RVI)W1 + (DMS)W2 + (RAQ)W3.$$

The weights of each component were determined using the analytical hierarchy process (AHP) method developed by Thomas L. Saaty (see Appendix A). For this, paired combinations were made between the three components using a pairwise comparison matrix (Table 1). For example, since the RVI is a quantitative value that reflects the loss or gain of cover in a given period, it was considered of greater relevance than the RAQ and DMS components. In the same way, among these last components, the DMS was considered of greater importance than RAQ. DMS is the result of the opinion of several experts (which includes years of experience) in comparison to RAQ, which takes information from a single field visit.

Table 1. Pairwise comparison matrix (PCM).

Criteria	Sub-Criteria	Number of Comparisons			Total	Weight
		1	2	3		
MCSI	(1) Remnant Vegetation Index		5	5	10	0.62
	(2) Delphi Method Survey.	1		3	4	0.25
	(3) Rapid Assessment Questionnaire.	1	1		2	0.13

2.3. *Remaining Vegetation Index (RVI)*

The value of RVI of each mangrove community was calculated by considering the vegetation cover obtained in 2018 as the present vegetation area (PVA), divided by the original vegetation area (OVA), which corresponded to the data of the year of 1973. The result was multiplied by 100 to obtain a comparable value on a scale of 0/100. This index was used for the first time in a case study in Colombia [18] following this formula:

$$\mathbf{RVI} = [(\mathrm{PVA})/(\mathrm{OVA})] \times 100.$$

We obtained the vegetation area from scanned aerial photographs and Landsat satellite images. We consulted CONABIO's database. We used the oldest image available from the sources mentioned above for the calculation of the RVI for each mangrove community. In this case, we obtained an aerial photograph from 1974 that captured mangroves, except those at Espíritu Santo Archipelago, in La Paz bay at the Autonomous University of Baja California Sur library's archive. We downloaded Sentinel images (10 m, 20 m, and 60 m, avoiding cloud interference) for May 2018 from the Earth Explorer platform (USGS) to calculate present vegetation cover for the calculation of the RVI.

To digitalize the aerial photographs, we scanned them with the highest available resolution (10,200 × 14,028 pixels). We georeferenced images using geomorphological land references by the control point method. Subsequently, we extracted the sections corresponding to mangroves on QGIS and obtained pixels (1 m × 1 m) by a resampling process. The resampling process did not allow for a higher pixel resolution; nevertheless, it provided better contrast between neighboring pixels. Therefore, observations allowed the precise definition of mangrove areas (Figure 3).

Together, the field data collection and the georeferenced aerial photographs allowed the confirmation of the presence of mangroves and the obtention of the polygons containing mangroves by the use of the on-screen scanning facility in the QGIS software. We transformed the satellite image from geographic coordinates to metrics. For optimal use, we created a composite of bands 4, 5, 3, and panchromatic (Band 8) to increase image resolution and facilitate vegetation recognition [19]. The generation of the base project in the QGIS platform integrated the resulting images in raster format. Once the properties of the images (pixel size, georeference) were validated, we calculated the mangrove cover for each of the 17 sites. We obtained 16 mangrove polygons for 1974 and 17 for 2018 using the manual digitizing technique which has been used by different authors [20–23] and estimated the vegetation cover area. We used the resulting areas to calculate the RVI.

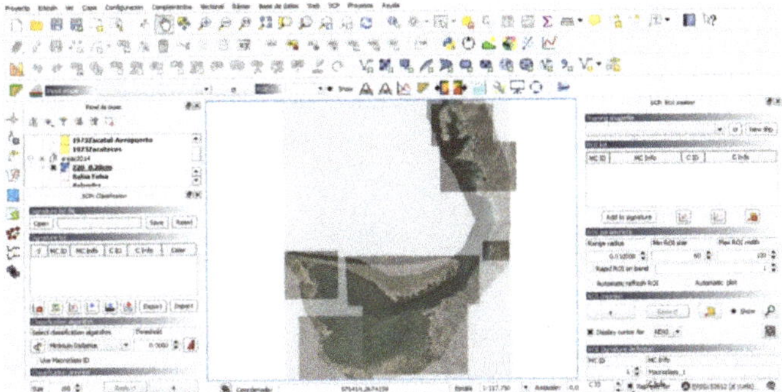

Figure 3. Treatment of spatial images in the QGIS software.

2.4. Delphi Method Survey

We applied interviews with regional mangrove experts following the Delphi method, which is a structured way to obtain information and knowledge on a particular topic [24]. This method provides both qualitative and quantitative data (see Appendix B), and it can be adapted for rapid assessments, such as the one implemented during this study [25]. We contacted a total of ten people, but only seven answered the survey. Of these, four were researchers, two worked for government agencies, and one collaborated with non-government organizations. We conducted interviews in person or remotely via electronic media such as Skype or video conference. The interview consisted mostly of open questions, as well as closed questions or fixed-alternatives. Interviewees answered the open question freely with no limit on time. The fixed-alternative questions were formulated to be answered in a scalar way using the Likert measurement tool (Table 2), which consists of obtaining a degree of conformity determined by a range of values. Table 3 shows the main question used in the survey.

Table 2. Remaining Vegetation Index obtained for each mangrove community.

Mangrove Community	Original Vegetation Area (m^2)	Present Vegetation Area (m^2)	Remaining Vegetation Index (Scores)
Balandra	266,044.01	268,577.08	**100.95**
Centenario-Chametla	54,213.98	53,657.47	98.97
Comitán	43,982.36	42,448.27	96.51
El Conchalito	217,785.87	192,957.92	88.59
El Merito	81,397.13	81,373.77	99.97
El Mogote	1,247,826.06	1,254,511.24	**100.53**
Enfermería	56,953.15	37,968.09	66.66
Eréndira	23,627.20	23,969.85	**101.45**
Espíritu Santo Archipelago	-	523,773. 69	N/A
Estero Bahía Falsa	47,744.44	44,573.39	93.35
Estero El Gato	47,916.84	45,429.86	94.80
La Paz-Aeropuerto	160,633.55	360,491.36	**224.41**
Palmira	14,172.06	12,154.07	85.76
Playa Pichilingue-Brujas	11,210.63	3,025.02	26.98
Salinas de Pichilingue	-	2,971.47	N/A
Unidad Pichilingue UABCS	63,284.70	50,329.61	79.52
Zacatecas	227,058.89	259,119.05	**114.11**

Note: Bold scores represent increases on mangrove cover.

Table 3. Key question applied to experts in the Delphi Method Survey component.

Based on your experience, what is the conservation status of the mangroves of the bay of La Paz?		
Mangrove Communities	**Conservation Status**	**I Don't know.**
Balandra	Bad 1 2 3 4 5 Good	

2.5. *Rapid Assessment Questionnaire*

The rapid evaluation is a reliable and timely estimation method, which allows an approximation of the magnitude and characteristics of a problem. It marks the line to define needs or tasks to consider during a subsequent evaluation [26]. This type of assessment provides complementary information to other sources, in a simple, fast, and flexible way. In the case of the mangroves of La Paz Bay, we visited 17 sites, which were selected according to the management plans of the protected area (Balandra) and Ramsar site (Humedales Mogote-Ensenada de La Paz No. 1816). To evaluate each of the mentioned mangroves, we created a rapid assessment questionnaire (*RAQ*) based on different surveys developed by academics and decision-makers from the region. The *RAQ* considered specific environmental indicators, divided thematically (water, air, soil, flora, fauna, and waste), and used qualitative indicators to assess impacts observed at each mangrove site during the field visits. The values of the *RAQ* ran from 0 to 1; the closer the value to 1, the more impacted the site was. We recorded our observations at the site, and photographic evidence is available from the authors upon request (see Appendix C).

2.6. *Application of the Integrative Mangrove Conservation Status Index*

We calculated the *MCSI* using the scores of each one of the components of the index, RVI, RAQ, and DMS, and following the formula:

$$\mathbf{MCSI} = (\text{RVI})(0.62) + (\text{RAQ})(0.25) + (\text{DMS})(0.13).$$

We classified mangrove sites depending on their MCSI score following an adapted classification of the IUCN Red List of Ecosystems (see Appendix D). This scale considers eight categories of risk for the earth's ecosystem (Figure 4). Three of them contemplate quantitative thresholds: critically endangered (CR), endangered (EN), and vulnerable (VU)—together, the IUCN describes these ecosystems as threatened. There are several qualitative categories to include: (1) ecosystems that fail to meet the quantitative criteria for the threatened ecosystem categories (NT, near threatened); (2) ecosystems that unambiguously meet none of the quantitative criteria (LC, least concern); (3) ecosystems with poor data (DD, data deficient); and (4) ecosystems that have not been assessed (NE, not evaluated). An additional category (CO, collapse) is assigned to ecosystems that have collapsed throughout their distribution, the analogue of the extinct (EX) category for species [27].

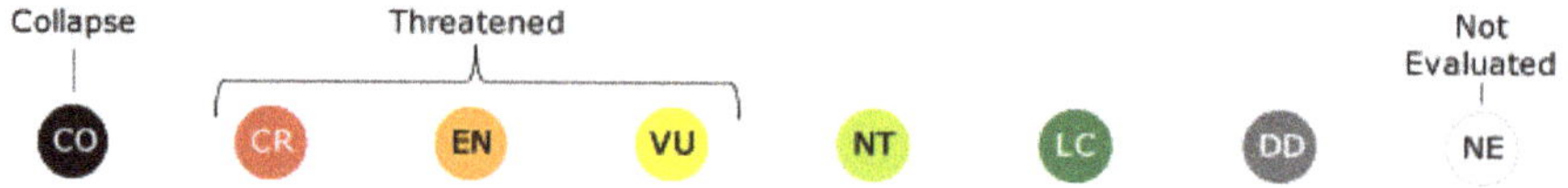

Figure 4. Categories of the IUCN Red List of Ecosystems. Source: IUCN, 2019.

3. Results

To calculate the proposed MCSI, we first estimated each of the components for the 17 mangroves from La Paz Bay. We summarize our findings in the following sections.

3.1. *Spatial Analysis and Remaining Vegetation Index (RVI)*

According to the RVI analysis, five communities presented an increase in mangrove cover (RVI > 100), ten showed losses of mangrove forest (RVI < 100), and two could not be analyzed because we were not able to obtain an original vegetation area. Playa Pichilingue-Brujas experienced the most significant losses of mangrove vegetation, and only 26.98% of the initial cover remained. Enfermería also presented a significant decrease in mangrove cover (44.44%).

3.2. Delphi Method Survey

We sent a total of 10 surveys to regional mangrove experts; however, just seven people replied. We interviewed them to complement their answers to the surveys (available from the authors upon request). We calculated the median for the score assigned by the experts to each one of the mangrove areas. Only two mangrove communities presented median scores of five, El Merito and Espíritu Santo Archipelago, which presents the best conservation status according to expert opinions (Table 4). Five mangrove areas scored four points for an acceptable conservation level. Moreover the mangroves that scored less than three, meant that they present a major deterioration.

Table 4. Values obtained from the Delphi Method Survey applied to experts.

Mangrove Community	Expert Opinion 1	Expert Opinion 2	Expert Opinion 3	Expert Opinion 4	Expert Opinion 5	Expert Opinion 6	Expert Opinion 7	Median
Balandra	4	3	4	4	5	4	4	4
Centenario-Chametla	2	2	3	2	4	2	2	2
Comitán	2	4	4	4	3	3	4	4
El Conchalito	3	2	3	3	4	2	3	3
El Merito	5	-	5	5	4	5	5	5
El Mogote	3	3	4	4	4	4	4	4
Enfermería	1	-	2	2	3	3	2	2
Eréndira	2	2	2	2	1	4	2	2
Espíritu Santo Archipelago	5	-	5	5	3	-	5	5
Estero Bahía Falsa	3	3	3	2	3	3	3	3
Estero El Gato	4	3	3	4	-	3	4	3
La Paz- Aeropuerto	4	3	4	3	5	4	4	4
Palmira	4	2	2	2	3	2	2	2
Playa Pichilingue-Brujas	2	2	1	1	4	2	2	2
Salinas de Pichilingue	-	-	3	3	-	-	-	-
Unidad Pichilingue UABCS	2	3	3	3	4	3	3	3
Zacatecas	4	-	5	4	4	4	5	4

3.3. Rapid Assessment Questionnaire

The mangroves in La Paz bay presented a varied rank of values for the RAQ. The La Paz-Aeropuerto, Palmira, and Playa Pichilingue–Brujas mangrove communities were the most impacted sites, with RAQ scores between 0.80 and 0.90. Seven mangroves showed medium levels of impact, with scores ranging between 0.50 and 0.70, and another seven sites presented fewer effects, with RAQ values from 0.26 to 0.49 (Table 5).

Table 5. Values obtained from the Rapid Assessment Questionnaire applied in field visits.

Mangrove Community	Rapid Assessment Questionnaires
Balandra	0.4944
Centenario-Chametla	0.6542
Comitán	0.2662
El Conchalito	0.5680
El Merito	0.3184
El Mogote	0.4486
Enfermería	0.5896
Eréndira	0.7260
Espíritu Santo Archipelago	0.2662
Estero Bahía Falsa	0.6382
Estero El Gato	0.5604
La Paz-Aeropuerto	0.8178
Palmira	0.8858
Playa Pichilingue-Brujas	0.8960
Salinas de Pichilingue	0.2884
Unidad Pichilingue UABCS	0.6504
Zacatecas	0.4026

3.4. Application of the Index

We multiplied the scores of the RVI, DMS, and RAQ, as described to calculate the MCSI. The El Merito mangrove community presented the highest MCSI of all the sites (95.54), followed by Balandra, Comitán, El Mogote, Estero El Gato, and Zacatecas, which scored values over 80; therefore, these five communities are of "least concern" in accordance with the IUCN Red List of Ecosystems. The MCSI values of another eight mangroves placed them as "near threatened". Otherwise, Enfermeria was classified as "vulnerable", with an MCSI score of 59.22, and Playa Pichilingue-Brujas was classified as "endangered" (MCSI score of 30.01), these last two mangroves were the worst evaluated. Lastly, two mangroves, Salinas de Pichilingue and Espíritu Santo Archipelago, lacked initial information on mangrove cover; therefore, the MCSI resulted in "data deficient" (Table 6).

Table 6. MCSI values for each mangrove community.

Mangrove Community	RVI Score	DM Score	RA Score	MCSI	IUCN Red List of Ecosystems
Balandra	57	23.2	7.0784	87.2784	Least Concern (LC)
Centenario-Chametla	57	11.6	4.8412	73.4412	Near Threatened (NT)
Comitán	57	17.4	10.2732	84.6732	Least Concern (LC)
Espíritu Santo Archipelago	-	29	10.2732	-	Data Deficient (DD)
El Conchalito	54.6231	17.4	6.048	78.0711	Near Threatened (NT)
El Merito	57	29	9.5424	95.5424	Least Concern (LC)
El Mogote	57	23.2	7.7196	87.9196	Least Concern (LC)
Enfermería	41.8836	11.6	5.7456	59.2292	Vulnerable (VU)
Eréndira	57	11.6	3.836	72.436	Near Threatened (NT)
Estero Bahía Falsa	57	17.4	5.0652	79.4652	Near Threatened (NT)
Estero El Gato	57	17.4	6.1544	80.5544	Near Threatened (NT)
La Paz-Aeropuerto	57	17.4	2.5508	76.9508	Near Threatened (NT)
Palmira	53.8593	11.6	1.5988	67.0581	Near Threatened (NT)
Playa Pichilingue-Brujas	16.9575	11.6	1.456	30.0135	Endangered (EN)
Salinas de Pichilingue	-	-	9.9624	-	Data Deficient (DD)
Pichilingue UABCS	45.5316	17.4	4.8944	67.826	Near Threatened (NT)
Zacatecas	57	23.2	8.3636	88.5636	Least Concern (LC)

4. Discussion

Various scientists have developed evaluation indices that can estimate the conservation status-health of mangroves [28–30]. However, many times the application of some indexes requires financial resources, specialized equipment, and experts, which are not always available. This occurs mainly in developing countries and at the same time in these sites information is required in an expedited manner for decision making. Therefore, we believe that the MCSI could help in the aforementioned scenario, which is so common in various countries in Latin America, Africa, and Asia.

Although this index is easy to apply and requires few financial resources, it is based on a combination of quantitative and qualitative data, which gives adequate support to the decisions generated from the results obtained. It combines mangrove cover (remnant vegetation index) with scientific experts' opinions (Delphi method survey), and perceived conservation status obtained during field visits (rapid assessment questionnaire) to classify mangroves in accordance with the IUCN Red List of Ecosystems. For that purpose, the MCSI used the analytical hierarchy process to define the weight of various indicators following a multicriteria method, as suggested by other authors [31].

Besides this, we can consider that the main finding of the construction and application of the MCSI index was that in general terms the mangrove communities located in the Bay of La Paz have an acceptable state of conservation. On the other hand, these results also indicate that despite being in the same area, mangroves could have very different conservation status.

The MCSI uses mangrove cover as a core indicator, which is also applied by the Mexican Mangrove Monitoring System (SMMM), but also is complemented by experts' opinions and rapid assessments at the mangrove communities. Still, different authors consider that cover is not sufficient to estimate the conservation [32,33], since it does not take into consideration other impacts on the ecosystems or the integrity of the ecological services they provided.

We complemented mangrove cover with the remaining vegetation index (RVI) because it estimates changes on vegetation cover in a specific period. The RVI determined an increase between 1974 and 2018 in the cover area of five mangrove communities in La Paz bay: Balandra, El Mogote, Eréndira, La Paz-Aeropuerto, and Zacatecas; however, the rest of the mangrove areas experienced a decrease in mangrove cover (Table 2). The most affected mangroves, Playa Pichilingue-Brujas and Enfermería, experimented significant human-induced impacts during the last decades, mainly the reduction of their connection to the sea by the construction of roads [34]. Mangroves present in La Paz bay facilitated the use of aerial and satellite imagery for the estimation of the RVI because they are adjacent to desertic areas, which showed high contrast with mangrove species. However, different authors consider that the cover estimations by this method are limited when mangroves are close to other types of forest or wetlands, and additional corrections are necessary [35,36].

The integration of data obtained by different methods has proven difficult [37]; therefore, this study used mixed methods to integrate qualitative and quantitative data, and most importantly, changes in mangrove cover with scientific expert's opinions. This last indicator is essential when the number of experts is limited, but their knowledge in the region is plenty. The information obtained by the DMS contrasted in some cases with the cover estimated by the RVI. In such cases, the scientific experts considered that some mangroves, e.g., El Mogote and Erendira, were in a poor state of conservation (Table 4). Still, those sites presented an increase in cover between 1974 and 2018, according to the RVI (Table 3). For those cases, the RAQ demonstrated the presence of some visible impacts, such as gray water inputs and modification of water circulation, which may be limiting the ecosystem services that mangroves should provide but increased their cover. Some of these impacts change over time; for example, solid waste was observed previously at Comitán but not registered during the RAQ thanks to a cleaning campaign that took place the day before the field visit. Therefore, the MCIS approach highlights the need to integrate expert opinions and field data.

The use of the MCSI scores to classify mangrove communities according to the IUCN Red List of Ecosystems is a novel approach. The analysis for 17 mangroves in La Paz bay resulted in five sites in the category of least concern (LC), and eight considered as near threatened (NT). On the other hand, one of them, the Enfermería mangrove area, is vulnerable (VU), and the Pichilingue-Brujas mangrove is endangered (EN). Finally, two mangroves, Salinas de Pichilingue and Espiritu Santo Archipelago, were classified as data deficient (DD). However, according to partial results, it is currently estimated that Salinas de Pichilingue and Espiritu Santo Archipelago have a good state of conservation, especially in the last case (Figure 5).

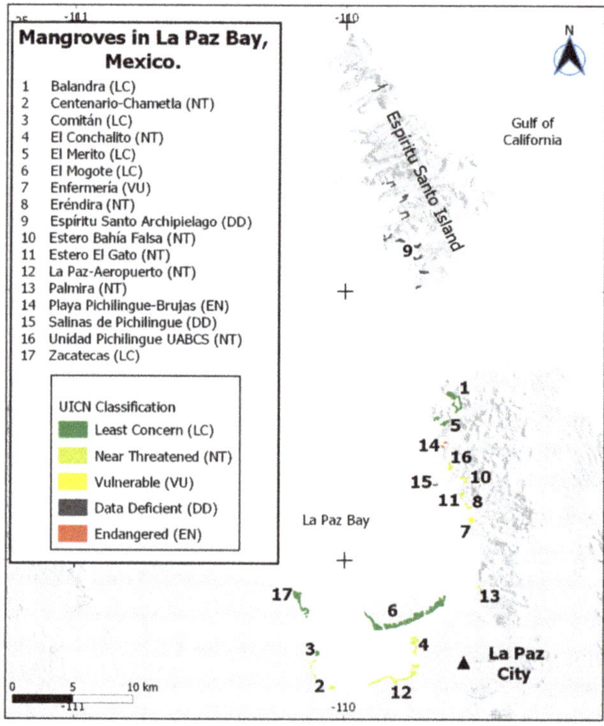

Figure 5. Classification of La Paz Bay mangroves, according to their conservation status.

Accordingly, the MCSI was adequate for the case study and helped to define conservation priorities for the mangroves in the region. The integrative nature of the index allowed for the identification of factors that negatively affect the conservation status of the mangroves, e.g., losses of vegetation, changes in water circulation, or solid waste presence. Then, it can be instrumental for the effective implementation of ecological restoration activities undertaken in areas [38,39]. The application of the MCSI by managers in this region may help to revert the condition of those mangrove areas that have suffered significant deterioration or to address other adverse factors threatening these ecosystems [40]. However, it is estimated that the urbanization caused by the tourism industry will be the most threatening factor for the conservation of mangroves in La Paz Bay, especially if it does not consider the applicable environmental regulations. Although several of the mangrove communities are small, due to their ecosystem services, these must be conserved [41].

Finally, it is concluded that the results on conservation status are more robust than those that include only spatial data and, by their integration into the IUCN Red List of Ecosystems, allow a direct comparison not only with other mangroves but also with different aquatic and terrestrial environments around the world. We recommend the application of the MCSI not only as a decision-making tool but also as an exploratory study; nonetheless, it is advisable to conduct follow up monitoring of quantitative ecological indicators to strengthen and provide feedback to update the MCSI.

5. Conclusions

The results of the present work constitute the first innovative use of the categorization of the Red List of Ecosystems by the IUCN for the mangroves communities. The MCSI integrates not only reliable quantitative indicators but also qualitative indicators to provide a more accurate conservation status for the mangrove communities; however, it is limited by the availability of data for some areas of interest. The multidisciplinary approach of the MCSI allows for its application at data deficient mangrove communities for an initial evaluation and guides future research efforts. The proposed MCSI index is a reliable tool for the management and conservation of mangrove communities; nonetheless, it could be improved as new methods to collect substantial data for additional indicators become available.

Author Contributions: Conceptualization, G.Á.-F., J.J.-M., O.A.-C., and G.H.-A.; Investigation, G.Á.-F.; Methodology, G.Á.-F. and P.C.-C.; Project administration, G.Á.-F. and J.J.-M.; Validation, G.Á.-F.; Writing—original draft, G.Á.-F. and G.H.-A.; Writing—Review and Editing, G.Á.-F., J.J.-M., P.C.-C., O.A.-C., J.M.L.-V. and G.H.-A. All authors have read and agreed to the published version of the manuscript.

Funding: The first author states support for this work was provided by the National Council of Science and Technology of Mexico (CONACYT) through a scholarship for his Ph.D. studies (CVU: 471803).

Acknowledgments: This paper is based partially on Ávila-Flores, 2016 and other data collected for the doctoral thesis of the first author. We thank Candidate Pablo Hernández-Morales, for his advice on spatial analysis and the M.Sc. Liliana Paredes-Lozano for her support during fieldwork and for proofreading the article. Finally, we would like to thank the mangrove experts who participated in the Delphi Method Survey and the three reviewers whose commentaries and suggestions greatly improved to this paper.

Conflicts of Interest: The authors declare no conflict of interest.

Appendix A

The analytical hierarchy process (AHP) proposed by Thomas L. Saaty (1980) is a method to express the relative importance or dominance of individual element(s) over an assessment, such as criteria or characteristics useful, for example, for decision-making [42]. This technique organizes the evaluation elements into a hierarchy level, assigning numerical values, which gives mathematical support to the organization. The AHP uses pairwise comparison or paired ranking to compare the various elements of the analysis. This comparison is based on a numerical qualification, considering a prime number subjacent scale from 1 to 9 (Saaty scale). Data are entered into a squared matrix to give qualifications; therefore, the level of importance (Table A1).

Table A1. Saaty scale used in paired ranking.

Scale Values	Values Definition	Definition
1	Equal importance	An equal level of importance for both evaluation elements that are being compared
3	Somewhat more or weak importance	One element is slightly more important or relevant than the other
5	Much more or essential importance	One element is more important or essential than the other.
7	Very much or demonstrated importance	One element is much more important or relevant than the other
9	Extreme or absolute importance	One element is definitely more important than the other

Appendix B

The Delphi survey method is a research technique proposed by The RAND Corporation in the 1950s (Table A2). It is a group communication process [43], with reference to a collection of answers in order to obtain and improve information resulting from opinions made by experts in a particular theme [44]. This methodology is based on the application of a series of questions, all of which are submitted once or several times if necessary, with the objective to achieve feedback and enrich opinions [45]. The origin and number of these may come from either the academic sector, industrial, governmental, or civil society organizations. It should be mentioned that the expert selection must be done considering their previous experience and current work on the subject. Using these opinions, this method seeks to develop a previously non-existent or dispersed knowledge useful in the decision-making process in government, science, and industry areas [46].

Table A2. Delphi method process description (based on Mukherjee et al. 2014).

1. FIRST ROUND
2. 🡇
3. PREPARE QUESTIONNAIRE
4. 🡇
5. SELECT EXPERTS IN ORDER TO OBTAIN THEIR OPINION
6. 🡇
7. INVITE THEM TO PARTICIPATE ON DELPHI
8. 🡇
9. COLLECT ANSWERS, ANALYSE, AND PROVIDE FEEDBACK
10. 🡇
11. SECOND ROUND
12. 🡇
13. PREPARE AND SEND QUESTIONNAIRE
14. 🡇
15. COLLECT ANSWERS, ANALYSE, AND PROVIDE FEEDBACK
16. 🡇
17. REPEAT ROUNDS UNTIL CONSENSUS IS REACHED

Appendix C

Rapid assessment or rapid appraisal (RA) is a quick and accurate estimation method to collect in situ information [47]. This technique allows for an immediate approximation to the magnitude and characteristics of a particular situation. It is an approach for developing a preliminary, qualitative understanding [48]. RA marks the line to define needs or tasks to be carried out in posterior assessments (Figure A1); it provides supplemental information to other sources, in a simple, fast, and flexible way [49]. The RA method uses various tools to collect data; for example, the recording of observations in logbooks or through questionnaires, participatory surveys. All these methods have the purpose of giving fast and reliable results. RA's fundamental advantage is that it is itself "a phase or stage in a research process"; therefore, it allows for scientific rigor [50].

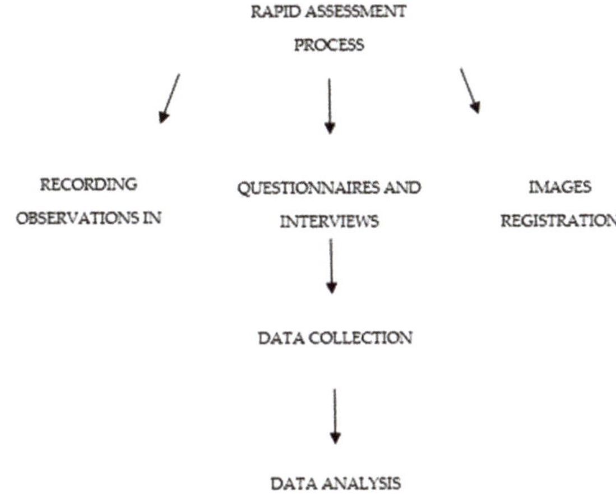

Figure A1. Rapid assessment process and elements.

Appendix D

The IUCN Red List of Ecosystems Categories and Criteria is a global standard for how we assess the status of ecosystems, applicable at local, national, regional, and global levels [51,52]. In this work, we artificially established the values to classify mangroves in each of the categories. Each criterion used for each category is described after the definitions mentioned below (Table A3).

Table A3. List of scores for each category of the MCSI Index. (Based on Keith et al., 2015).

IUCN Red List of Ecosystems	Score
Collapsed (CO)	0
Critically endangered (CR)	01–20
Endangered (EN)	21–40
Vulnerable (VU)	41–60
Near Threatened (NT)	61–80
Least Concern (LC)	81–100
Data Deficient (DD)	N/A
Not Evaluated (NE)	N/A

- Collapsed (CO)

"An ecosystem is Collapsed when it is virtually certain that its defining biotic or abiotic features are lost from all occurrences, and the characteristic native biota are no longer sustained. Collapse may occur when most of the diagnostic components of the characteristic native biota are lost from the system, or when functional components (biota that performs key roles in ecosystem organization) are greatly reduced in abundance and lose the ability to recruit" (IUCN, 2019). In this category, the entire vegetation cover of the mangrove community has been destroyed.

- Critically Endangered (CR)

"An ecosystem is Critically Endangered when the best available evidence indicates that it meets any of the criteria A to E for Critically Endangered. It is therefore considered to be at an extremely high risk of collapse" (IUCN, 2019). In this category, the index score obtained by the mangrove community is less than 20 but higher than zero.

- Endangered (EN)

"An ecosystem is Endangered when the best available evidence indicates that it meets any of the criteria A to E for Endangered. It is therefore considered to be at a very high risk of collapse" (IUCN, 2019). In this category, the index score obtained by the mangrove community is less than 40 but higher than 20.

- Vulnerable (VU)

"An ecosystem is Vulnerable when the best available evidence indicates that it meets any of the criteria A to E for Vulnerable. It is therefore considered to be at a high risk of collapse" (IUCN, 2019). In this category, the index score obtained by the mangrove community is less than 60 but higher than 40.

- Near Threatened (NT)

"An ecosystem is Near Threatened when it has been evaluated against the criteria but does not qualify for Critically Endangered, Endangered or Vulnerable now, but is close to qualifying for or is likely to qualify for a threatened category in the near future" (IUCN, 2019). In this category, the index score obtained by the mangrove community is less than 80 but higher than 60.

- Least Concern (LC)

"An ecosystem is Least Concern when it has been evaluated against the criteria and does not qualify for Critically Endangered, Endangered, Vulnerable or Near Threatened. Widely distributed and relatively undegraded ecosystems are included in this category" (IUCN, 2019). In this category, the index score obtained by the mangrove community is equal to or less than 100 but higher than 80.

- Data Deficient (DD)

"An ecosystem is Data Deficient when there is inadequate information to make a direct, or indirect, assessment of its risk of collapse based on decline in distribution, disruption of ecological function or degradation of the physical environment. Data Deficient is not a category of threat, and does not imply any level of collapse risk" (IUCN, 2019). In this category, some of the information considered for the application of the index in a particular mangrove community was not available.

- Not Evaluated (NE)

"An ecosystem is Not Evaluated when it is has not yet been assessed against the criteria" (IUCN, 2019). In this category, it means that it was not evaluated in any way within the three components considered within the index.

References

1. Spalding, M.D.; Kainuma, M.; Collins, L. *World atlas of Mangroves*; Earthscan: London, UK, 2010; pp. 24–25.
2. Van Bochove, J.; Sullivan, E.; Nakamura, E. *The Importance of Mangrove to People: A Call to Action*; United Nations Environment Programme, World Conservation Monitoring Centre: Cambridge, UK, 2014; pp. 43–45.
3. Vo, Q.T.; Kuenzer, C.; Vo, Q.M.; Moder, F.; Oppelt, N. Review of valuation methods for mangrove ecosystem services. *Ecol. Indic.* **2012**, *23*, 431–446. [CrossRef]
4. Atkinson, S.C.; Jupiter, S.D.; Adams, V.M.; Ingram, J.C.; Narayan, S.; Klein, C.J.; Possingham, H.P. Prioritising mangrove ecosystem services results in spatially variable management priorities. *PLoS ONE* **2016**, *11*, e0151992. [CrossRef] [PubMed]
5. Kadykalo, A.N.; López-Rodríguez, M.D.; Ainscough, J.; Droste, N.; Ryu, H.; Ávila-Flores, G.; Harmáčková, Z. Disentangling 'ecosystem services' and 'nature's contributions to people'. *Ecosyst. People* **2019**, *15*, 269–287. [CrossRef]
6. Roldán, V.A.; Galván, D.E.; Lopes, P.F.; López, J.; Bellamy, A.S.; Gallego, F.; Barriga, A.M. Are we seeing the whole picture in land-sea systems? Opportunities and challenges for operationalizing the ES concept. *Ecosyst. Serv.* **2019**, *38*, 100966. [CrossRef]
7. Sandilyan, S.; Kathiresan, K. Mangrove conservation: A global perspective. *Biodivers. Conserv.* **2012**, *21*, 3523–3542. [CrossRef]
8. Romañach, S.S.; DeAngelis, D.L.; Koh, H.L.; Li, Y.; Teh, S.Y.; Barizan, R.S.R.; Zhai, L. Conservation and restoration of mangroves: Global status, perspectives, and prognosis. *Ocean Coast. Manag.* **2018**, *154*, 72–82. [CrossRef]
9. Verdugo, F.F.; Casasola, P.M.; Hernández, C.M.A.; Rosas, H.L.; Pardo, D.B.; Bello, A.C.T. La topografía y el hidroperiodo: Dos factores que condicionan la restauración de los humedales costeros. *Boletín Soc. Botánica México* **2007**, *80*, 33–47.
10. FAO. *The World's Mangroves 1980–2005*; FAO Forestry Paper 153; FAO: Rome, Italy, 2007; p. 77.
11. Duke, N.C.; Meynecke, J.O.; Dittmann, S.; Ellison, A.M.; Anger, K.; Berger, U.; Koedam, N. A world without mangroves? *Science* **2007**, *317*, 41–42. [CrossRef]
12. Giri, C.; Ochieng, E.; Tieszen, L.L.; Zhu, Z.; Singh, A.; Loveland, T.; Duke, N. Status and distribution of mangrove forests of the world using earth observation satellite data. *Glob. Ecol. Biogeogr.* **2011**, *20*, 154–159. [CrossRef]
13. Sitios de Manglar con Relevancia Biológica y Con Necesidades de Rehabilitación Ecológica. Available online: http://www.biodiversidad.gob.mx/ecosistemas/manglares2013/sitioPacNorte.html (accessed on 20 September 2019).
14. Dávalos-Sotelo, R. El papel de la investigación científica en la creación de las áreas naturales protegidas. *Madera y Bosques* **2016**, *22*, 7–13. [CrossRef]
15. Ávila-Flores, G.; Juárez-Mancilla, J.; Hinojosa-Arango, G.O.; Covarrubias, A. The Use of the DPSIR Framework to estimate Impacts of Urbanization on Mangroves: A Case Study from La Paz, Baja California Sur, Mexico. In *WIT Transactions on Ecology and the Environment*; Sustainable City XII; Brebbia, C.A., Sendra, J.J., Eds.; WIT Press: Southampton, UK, 2017; Volume 223, pp. 459–469.
16. Places to Go in 2020. The New York Times. Available online: https://www.nytimes.com/interactive/2020/travel/places-to-visit.html (accessed on 11 January 2020).
17. González-Baheza, A.; Arizpe, O. Vulnerability assessment for supporting sustainable coastal city development: A case study of La Paz, Mexico. *Clim. Dev.* **2018**, *10*, 552–565. [CrossRef]
18. Márquez, G. Colombia: Ambiente, pobreza, violencia. *Fermentum Rev. Venez. Sociol. Antropol.* **2003**, *13*, 25–37.
19. Chauvaud, S.; Bouchon, C.; Manieres, R. Remote sensing techniques adapted to high resolution mapping of tropical coastal marine ecosystems (coral reefs, seagrass beds and mangrove). *Int. J. Remote Sens.* **1998**, *19*, 3625–3639. [CrossRef]
20. Benfield, S.L.; Guzman, H.M.; Mair, J.M. Temporal mangrove dynamics in relation to coastal development in Pacific Panama. *J. Environ. Manag.* **2005**, *76*, 263–276. [CrossRef] [PubMed]
21. Fromard, F.; Vega, C.; Proisy, C. Half a century of dynamic coastal change affecting mangrove shorelines of French Guiana. A case study based on remote sensing data analyses and field surveys. *Mar. Geol.* **2004**, *208*, 265–280. [CrossRef]

22. Murray, M.R.; Zisman, S.A.; Furley, P.A.; Munro, D.M.; Gibson, J.; Ratter, J.; Bridgewater, S.; Minty, C.D.; Place, C.J. The mangroves of Belize. Part 1. Distribution, composition and classification. *For. Ecol. Manag.* **2003**, *174*, 265–279. [CrossRef]
23. Cissell, J.R.; Delgado, A.M.; Sweetman, B.M.; Steinberg, M.K. Monitoring mangrove forest dynamics in Campeche, Mexico, using Landsat satellite data. *Remote Sens. Appl.* **2018**, *9*, 60–68. [CrossRef]
24. Okoli, C.; Pawlowski, S.D. The Delphi method as a research tool: An example, design considerations and applications. *Inf. Manag.* **2004**, *42*, 15–29. [CrossRef]
25. Hasson, F.; Keeney, S.; McKenna, H. Research guidelines for the Delphi survey technique. *J. Adv. Nurs.* **2000**, *32*, 1008–1015.
26. Kelly, C. *Guidelines for Rapid Environmental Impact Assessment in Disasters*; Benfield Hazard Research Centre, University of London: London, UK, 2005.
27. Rodríguez, J.P.; Rodríguez-Clark, K.M.; Baillie, J.E.; Ash, N.; Benson, J.; Boucher, T.; Keith, D.A. Establishing IUCN red list criteria for threatened ecosystems. *Conserv. Biol.* **2011**, *25*, 21–29. [CrossRef]
28. Kovacs, J.M.; Vandenberg, C.V.; Wang, J.; Flores-Verdugo, F. The use of multipolarized spaceborne SAR backscatter for monitoring the health of a degraded mangrove forest. *J. Coast. Res.* **2008**, *24*, 248–254. [CrossRef]
29. Chellamani, P.; Singh, C.P.; Panigrahy, S. Assessment of the health status of Indian mangrove ecosystems using multi temporal remote sensing data. *Trop. Ecol.* **2014**, *55*, 245–253.
30. Faridah-Hanum, I.; Yusoff, F.M.; Fitrianto, A.; Ainuddin, N.A.; Gandaseca, S.; Zaiton, S.; Shamsuddin, I. Development of a comprehensive mangrove quality index (MQI) in Matang Mangrove: Assessing mangrove ecosystem health. *Ecol. Indic.* **2019**, *102*, 103–117. [CrossRef]
31. Wątróbski, J.; Ziemba, E.; Karczmarczyk, A.; Jankowski, J. An index to measure the sustainable information society: The Polish households case. *Sustainability* **2018**, *10*, 3223. [CrossRef]
32. Chowdhury, A.; Maiti, S.K. Assessing the ecological health risk in a conserved mangrove ecosystem due to heavy metal pollution: A case study from Sundarbans Biosphere Reserve, India. *Hum. Ecol. Risk Assess.* **2016**, *22*, 1519–1541. [CrossRef]
33. Aguirre-Rubí, J.R.; Ortiz-Zarragoitia, M.; Izagirre, U.; Etxebarria, N.; Espinoza, F.; Marigómez, I. Prospective biomonitor and sentinel bivalve species for pollution monitoring and ecosystem health disturbance assessment in mangrove–lined Nicaraguan coasts. *Sci. Total Environ.* **2019**, *649*, 186–200. [CrossRef] [PubMed]
34. Ávila-Flores, G.; Riosmena-Rodríguez, R. Integrated environmental assessment and scenarios of mangrove community in the Gulf of California case study 'Manglar Enfermería'. In *The Arid Mangrove Forest from Baja California Peninsula 2*; Nova Publishers: New York, NY, USA, 2016; Volume 2, pp. 131–150.
35. Flores-de-Santiago, F.; Kovacs, J.M.; Flores-Verdugo, F. Discrimination of 3 dominant mangrove species from the Pacific coast of Mexico by spectroscopy on intact leaves. *Cienc. Mar.* **2018**, *44*, 185–202. [CrossRef]
36. Pham, T.D.; Yokoya, N.; Bui, D.T.; Yoshino, K.; Friess, D.A. Remote sensing approaches for monitoring mangrove species, structure, and biomass: Opportunities and challenges. *Remote Sens.* **2019**, *11*, 230. [CrossRef]
37. Wątróbski, J.; Jankowski, J.; Ziemba, P.; Karczmarczyk, A.; Zioło, M. Generalised framework for multi-criteria method selection. *Omega* **2019**, *86*, 107–124. [CrossRef]
38. López-Portillo, J.; Lewis, R.R.; Saenger, P.; Rovai, A.; Koedam, N.; Dahdouh-Guebas, F.; Rivera-Monroy, V.H. Mangrove forest restoration and rehabilitation. In *Mangrove Ecosystems: A Global Biogeographic Perspective*; Springer: Cham, Germany, 2017; pp. 301–345.
39. Reyes, M.A.; Tovilla, C. Restauración de áreas alteradas de manglar con Rhizophora mangle en la Costa de Chiapas. *Madera Bosques* **2002**, *8*, 103–114. [CrossRef]
40. Avila-Flores, G. Diagnosis of the conservation status of mangroves in the city of La Paz. Master's Thesis, Baja California Sur, Mexico, February 2016.
41. Curnick, D.; Pettorelli, N.; Amir, A.; Balke, T.; Barbier, E.; Crooks, S.; Quarto, A. The value of small mangrove patches. *Science* **2019**, *363*, 239. [PubMed]
42. Wind, Y.; Saaty, T.L. Marketing applications of the analytic hierarchy process. *Manag. Sci.* **1980**, *26*, 641–658. [CrossRef]
43. Hsu, C.C.; Sandford, B.A. The Delphi technique: Making sense of consensus. *Pract. Assess. Res. Eval.* **2007**, *12*, 1–8.
44. Gordon, T.J. The Delphi method. *Futures Res. Methodol.* **1994**, *2*, 1–30.

45. Mukherjee, N.; Huge, J.; Sutherland, W.J.; McNeill, J.; Van Opstal, M.; Dahdouh-Guebas, F.; Koedam, N. The Delphi technique in ecology and biological conservation: Applications and guidelines. *Methods Ecol. Evol.* **2015**, *6*, 1097–1109. [CrossRef]
46. Renzi, A.B.; Freitas, S. The Delphi method for future scenarios construction. *Procedia Manuf.* **2015**, *3*, 5785–5791. [CrossRef]
47. Beebe, J. *Basic Concepts and Techniques of Rapid Appraisal*; Human organization: New York, NY, USA, 1995; pp. 2–51.
48. Beebe, J. *Rapid Assessment Process: An Introduction*; AltaMira Press: Lanham, MD, USA, 2001.
49. Hjortsø, C.N.; Christensen, S.M.; Tarp, P. Rapid stakeholder and conflict assessment for natural resource management using cognitive mapping: The case of Damdoi Forest Enterprise, Vietnam. *Agric. Hum. Values* **2005**, *22*, 149–167. [CrossRef]
50. Pedersen, D.; Scrimshaw, N.S.; Gleason, G.R. *Rapid Assessment Procedures: Qualitative Methodologies for Planning and Evaluation of Health Related Programmes*; INFDC: Boston, MA, USA, 1992.
51. Keith, D.A.; Rodríguez, J.P.; Brooks, T.M.; Burgman, M.A.; Barrow, E.G.; Bland, L.; Miller, R.M. The IUCN red list of ecosystems: Motivations, challenges, and applications. *Conserv. Lett.* **2015**, *8*, 214–226. [CrossRef]
52. IUCN. Red List of Ecosystems Web. Available online: https://iucnrle.org/ (accessed on 25 October 2019).

Article

Identifying Key Knowledge Gaps to Better Protect Biodiversity and Simultaneously Secure Livelihoods in a Priority Conservation Area

Anke S. K. Frank * and Livia Schäffler

Zoological Research Museum Alexander Koenig – Leibniz Institute for Animal Biodiversity, Adenauerallee 160, 53113 Bonn, Germany; l.schaeffler@leibniz-zfmk.de

* Correspondence: a.frank@leibniz-zfmk.de; Tel.: +49-228-9122-247

Received: 23 September 2019; Accepted: 11 October 2019; Published: 15 October 2019

Abstract: Global agreements like the Sustainable Development Goals (SDGs) and Achi Biodiversity Targets (ABTs) aim to secure human well-being and to protect biodiversity, but little progress has been made in reaching these aims. The key role of biodiversity in securing human well-being is rarely considered a priority – instead short-term economic profits benefiting a few are prioritized. Particularly where local livelihoods rely on resources of protected areas for immediate survival, top-down enforced biodiversity conservation often increases social inequality, hunger and poverty and thus regularly fails. Identifying key knowledge gaps helps to adjust political priority setting and investment strategies to assess conservation threats and improve natural resource management. Since acting usually occurs at a local or regional scale, we focused on a priority conservation area in one of the world's poorest countries — the dry deciduous forests of western Madagascar. We aimed to identify key knowledge gaps in this area which need to be filled to better protect biodiversity and simultaneously ensure well-being of the local poor. We consulted 51 predominantly Malagasy experts using questionnaires. These questionnaires listed 71 knowledge gaps we collated from the literature which the experts were asked to rank by importance. Experts were encouraged to list additional knowledge gaps. Averaging the scores of all experts, we identified the top 10 knowledge gaps. Two political knowledge gaps addressing the need to determine strategies which improve law enforcement and reduce corruption ranked highest, followed by an ecological one concerning appropriate restoration and a socio–economic one regarding economic benefits locals gain from biodiversity. The general knowledge gap perceived as most important addressed strategies for long-term funding. Only one additional knowledge gap was identified: the impact of climate change-driven human migration from southwestern to central western Madagascar on socio–economic problems and its impacts on natural resources We linked the identified top 10 knowledge gaps as well as the additional knowledge gap suggested by experts to the SDGs, ABTs and 2 °C target of the Paris Climate Agreement, and discussed why these gaps were considered a priority. This research highlights important ecological, socio–economic and political research priorities and provides guidelines for policy makers and funding organizations.

Keywords: Aichi Biodiversity Targets; biodiversity conservation; dry deciduous forest; human well-being; Madagascar; Paris Climate Agreement; Sustainable Development Goals

---

1. Introduction

1.1. *Theoretical Background*

The Sustainable Development Goals (SDGs), the Aichi Biodiversity Targets (ABTs) and the 2 °C target of the Paris Climate Agreement all aim to secure human well-being in a sustainable world [1–3]. Even though human wellbeing essentially relies on the preservation of biodiversity to ensure ecosystem functioning [4,5], current global conservation efforts fail to halt biodiversity decline, which occurs at an unprecedented rate and will continue to do so due to ongoing as well as new threats [6–8]. This has extensive negative effects on economy and society due to ecosystem service losses [9] and clearly demonstrates that the key role of biodiversity for human well-being is not easily recognized by politicians aiming to achieve these global targets. Despite recent efforts to uncover the synergies and trade-offs between the goals and targets of these conventions [10–13], there remains a high risk that nations will cherry-pick a few goals suiting their priorities and fail to tackle those harder to accomplish [14]. Similarly, the ABTs have been criticized for their conflicting interests and lack of indicators, particularly concerning the drivers of biodiversity loss [15,16]. While new indicators are constantly being developed and others improved or upgraded regarding their availability for more countries [17], essential knowledge gaps remain, particularly concerning data from developing countries [18]. Identifying key knowledge gaps is important for assessing biodiversity threats [19] as well as for improving monitoring, management and investment strategies [20,21]. Approaching the end of the United Nations Decade on Biodiversity [22], we aim to identify key knowledge gaps concerning biodiversity loss in a particular priority conservation area as an example for many other priority conservation areas to be used by practitioners to streamline funds, resources and efforts to tackle the ongoing biodiversity crisis while simultaneously secure livelihoods.

Most people's well-being predominantly depends on ecosystem services provided by terrestrial biodiversity. However, there are trade-offs between protection of life on land and human wellbeing. Strict area protection for conservation has frequently cut off the local poor from essential resources and thus led to famine and increased social inequality threatening their survival [23–25]. We therefore have focused our knowledge gap search on the protection of "Life on land" (SDG 15) and major trade-offs of this central goal with SDG 2 (zero hunger—directly connected to food security through agricultural land use), and SDG 10 (reduced inequalities) (Figure 1). The predominant trade-off between SDG 2 and SDG 15 is the expansion of agricultural areas to reduce hunger (SDG 2) resulting in competition for land with SDG 15 aiming to protect natural ecosystems like forests and their biodiversity [15]. Progress in achieving SDG 15 has often reduced that of SDG 10 [26–28]. Social and material inequality have been shown to be harmful to the environment and thus to people's health and well-being [29,30].

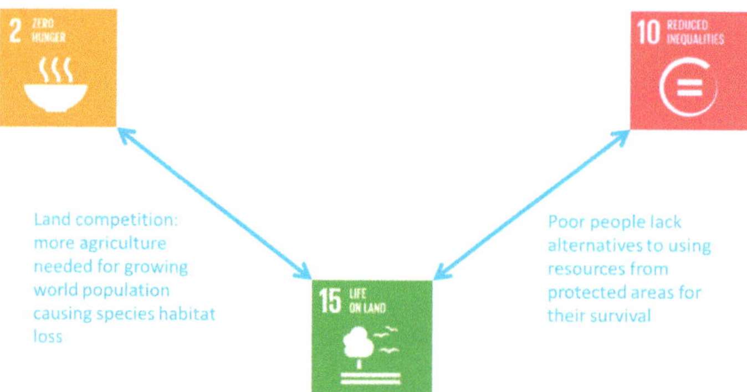

Figure 1. Trade-offs between Sustainable Development Goal (SDG) 2, SDG 10 and SDG 15.

1.2. Regional Focus Area: the Dry Deciduous Forests of Western Madagascar

As countries and regions differ in their demographics, geographies and governance [31], and are differently affected by climate change and biodiversity threats [32], we selected one regional example which promises high conservation payoff (in terms of biodiversity protection): the dry deciduous forest in western Madagascar. Madagascar is one of many developing countries rich in biodiversity and natural resources, but economically highly disadvantaged [33]. It is one of the poorest countries on Earth, where malnutrition is prevalent and where about 80% of the population lives below the poverty line [33,34]. Despite a growth in GDP, poverty has been increasing [35] and over half of the population faces food insecurity [36]. Madagascar's population has increased by 400% over the last five decades to about 23.5 million [37,38]. This massive population growth, demanding an increasing share of land for agriculture, can be considered the major indirect driver of dry forest loss in western Madagascar [39]. However, political turmoil, insecurity and corruption are also major issues in Madagascar. Environmental crimes encompass illegal land clearance for large-scale agricultural expansion, illegal timber and wildlife trade, and mining [40]. Political crises occur on an almost decadal basis, preventing transparency and accountability of governmental actions [33,41,42]. Currently, Madagascar is listed at fifth position in the global ranking of increased risk of notable changes to the Global Peace Index [43], dropped severely in the Rule of Law Index [40], and only ranks 155th out of 180 nations in the International Corruption Perception Index [44]. Moreover, Madagascar ranks in the bottom 10 of 51 African countries assessed for their performance in reaching the SDGs [45]. Due to successful international awareness-raising of the uniqueness of and threats to Madagascar's biodiversity, the country has seen a quadrupling in protected areas since 2003, with about 10% of the country being protected for conservation [46,47]. However, like in many tropical countries, conservation has focused on rainforests in the east of Madagascar, while the diverse dry forests in the west and south have rather been neglected [39,48,49]. But even where protected areas had been declared, these largely failed to prevent forest loss and degradation [33,50,51] due to a lack of law enforcement and high levels of corruption [33,40,42,47]. At least 13 protected areas are considered mere "paper parks" as they are totally devoid of management [47]. Generally, many knowledge gaps remain in Madagascar for achieving fair and equitable biodiversity conservation [52]. We chose the highly threatened dry deciduous forest of western Madagascar because forest losses predominantly caused by slash-and-burn agriculture and illegal logging are still dramatic there despite some national and international conservation attempts. Almost three quarters of the population live from subsistence farming practicing slash-and-burn agriculture [46,53] due to lack of alternatives but also because this farming practice is part of the people's cultural identity [54,55]. Slash-and-burn agriculture is conducted in two stages: during the dry season (June to September), woody undergrowth is cut and stacked around trees; at the beginning of the growth season (October) these piles of undergrowth are ignited, resulting in the destruction of all vegetation except for a few dead blackened tree trunks [56]. This form of agriculture requires little labor and—because of the ashes—requires no addition of nutrients for two to five years [57]. Afterwards, the land needs to be left fallow for several years (at least 20 years within the dry forest—see [58]). A reduced or lacking fallow period results in severe nutrient loss and too frequent burning favors the establishment of introduced and invasive species, preventing native species regeneration [58,59]. A growing need for agricultural land of the rapidly increasing population hinders sustainable management that would allow for sufficient regeneration time of soils in agricultural fields. Thus, the rate of turning primary forests into agricultural fields by slash and burn practices to generate fertile farmland is increasing. About 40% of the forest have been lost since 1970 [39]. Fragmentation has been immense, so that few areas of primary forest larger than 800 ha remain – too small to contain viable populations of many species like larger lemurs [60]. Fragmentation also hinders animals to disperse and migrate to cope with climate change conditions [61,62]. The largest remaining area of dry deciduous forest occurs in central-western Madagascar [63]: Menabe Central ranks among the hottest biodiversity hotspots in the world [60], particularly due to its exceptionally high rate of endemism and intense anthropogenic threats [64]. While the total number of species is lower in the dry than in the humid forests, species richness is exceptionally high by global comparison with other dry forests [39]. Madagascar's dry forests harbor several locally endemic vertebrate species such as Madame Berthe's mouse lemur (*Microcebus berthae*), the giant jumping rat (*Hypogeomys antimena*), the narrow-striped mongoose (*Mungotictis decemlineata*), and plenty of other endangered species [65–68]. *M. berthae*, the smallest primate in the world, is particularly vulnerable to anthropogenic disturbances and restricted to core habitats of this biome [66,69,70]. Without immediate protection of its habitat, this species will likely be extinct by 2050 [63], just as many other vertebrate species of lemurs, rodents or tenrecs have already been driven to extinction [39]. This is of concern due to the roles these species play for ecosystem functioning, its regeneration abilities and hence long-term persistence which for example depends on the seed dispersing role of lemurs [71]. Western Madagascar's dry deciduous forests are also considered of high conservation importance by a range of other approaches identifying priority conservation areas, for example the "Global 200" ecoregions [72] due to many endemic and endangered species and the Key Biodiversity Areas with a high biological value and intense anthropogenic pressure [73,74].

With respect to the people inhabiting this priority conservation area, social inequality is a huge issue hampering sustainable resource utilization as well as human well-being. Despite ambitious intentions, limited understanding of social–economic and ecological contexts prevented the realization of an effective protected area network [47]. For example, in spite of plans to involve communities in decision-making processes under the Durban Vision aiming to triple the amount of protected areas [47], conservation decisions have mostly been top-down enforced, sharply restricting or banning the use of local resources on which the local population relied [48,75]. This led to increased hunger, poverty and inequality of the already poor population as well as widespread criticism of the prioritization of the survival of lemurs over the survival of people [ibid.]. Due to funding, personnel and time restrictions of this project, we were unable to involve local community members in the collation and ranking of knowledge gaps in this study and therefore had to rely on predominantly academic experts for our knowledge gap ranking. To increase the chance of locals' needs to be considered in the ranking of knowledge gaps, we asked our participants whether they had work experience with local people.

A range of attempts have been made to identify knowledge gaps or important research questions, e.g., in the reports of the Intergovernmental Platform for Biodiversity and Ecosystem Services (IPBES) or by the yearly scan for the most important 100 ecological research questions at the time (see [21] at the global scale, and [76] at the national (UK) scale). However, acting usually occurs at the regional or local level and thus considering the specific local or regional conditions and needs of people at this level is important [77]. By choosing a regional example (i.e., the western dry deciduous forest of Madagascar) and conducting a consultation exercise using questionnaires with professionals from various fields (including subsistence farming and forest biodiversity, human development, governance, etc.), we followed an inclusive approach to "overcome the limitations of a consultation exercise of global aspirations" (Oldekop et al. 2016).

1.3. Aims of Our Study

We aimed to provide practitioners, politicians and funding bodies with a top 10 list of knowledge gaps concerning the slash-and-burn problematic in western Madagascar which need to be filled most urgently. This top 10 list is intended to be used as a guideline for decision making in research efforts and funding distribution. We undertook the following steps to pinpoint these key knowledge gaps:

(1) Identify knowledge gaps concerning biodiversity conservation with a focus on SDGs 2, 10 and 15 at the global scale to spot knowledge gaps relevant to the slash-and-burn problematic in western Madagascar.
(2) List knowledge gaps relevant for the slash-and-burn problematic threatening the dry deciduous forest in western Madagascar and sort into categories (ecological, socio–economic, political and general knowledge gaps).
(3) Identify the top 10 knowledge gaps which need to be tackled most urgently to enhance biodiversity conservation success while securing local livelihoods in our focal region (dry deciduous forests of western Madagascar) by using the help of experts via questionnaires.
(4) Identify links of these key knowledge gaps to our focal and any other goal and target of the SDGs and ABTs as well as the 2 °C target of the Paris Climate Agreement as these goals and targets are highly interrelated [10,12,78].
(5) Discuss previous attempts to fill some of these knowledge gaps in Madagascar (e.g., by checking Madagascar's National Biodiversity Strategy and Action Plan (NBSAP).

In this paper, we will present and discuss the expert-identified top 10 priority knowledge gaps, link them to the SDGs, ABTs and 2 °C target of the Paris Climate Agreement and examine the effects of experts' backgrounds on the ranking to check whether a particular type of characteristic (e.g., profession, age, gender) influenced the ranking scores.

2. Materials and Methods

2.1. Knowledge Gap Identification at the Global Scale and Sorting into Categories

In a first literature-based step, knowledge gaps related to the trade-offs between SDG 15 (life on land) and SDG 2 (zero hunger) as well as SDG 10 (reduced inequality) were identified at a global scale to spot knowledge gaps that are not directly addressed in the regional literature dealing with the slash-and-burn problematic in western Madagascar. We conducted a modified "snowball principle" based literature search using any paper which yielded relevant knowledge gaps to identify further relevant papers, as well as to check for subsequent citing publications [79,80]. Papers studied for knowledge gaps were restricted to those from the last 10 years, as recommended for a snowball literature search [81]. Further knowledge gaps were added by the authors based on their own professional experience.

Knowledge gaps relevant for the slash-and-burn problematic in dry deciduous forests of western Madagascar were listed (n = 71). Most of the literature sourced to select relevant knowledge gaps for this area concerned the core area of Menabe Central. Again, we used the snowball principle for our literature search as well as judgement of relevance by author LS who has extensive work experience in central western Madagascar. We separated these 71 knowledge gaps into ecological (n = 26), socio–economic (n = 24), and political categories (n = 14), because these categories relate to the different spheres of sustainability (i.e., economic, social, environmental and political, see [82] and because it is necessary to identify solutions that have "traction in the social, economic, and political arenas in which conservation action must take place" [83]). In addition, we listed the category "general" (n = 7)—comprising knowledge gaps that apply to any of the other categories (see Questionnaire in the Supplementary Material S1).

2.2. Identification of Key Knowledge Gaps through Expert Ranking

Knowledge gaps relevant for the slash-and-burn problematic in western Madagascar were listed in a questionnaire (see S1) designed following the guidelines of McLafferty [84]. To keep the questionnaire short, we combined some knowledge gaps which were closely linked. After discussion of the questionnaires with some experts of our focal area in Madagascar, the questionnaires were refined and then sent to representatives from universities and other research institutions as well as NGOs that have a long professional experience in Madagascar (via Email or LinkedIn). For time and logistic reasons, we could not include local people in the consultation process. However, to make sure that their perspective was represented, we have consulted experts that had been predominately working with rural people. We used snowball sampling [85] to recruit further experts by asking those contacted to spread the questionnaires further via their own personal networks.

The questionnaires consisted of two parts. Part I asked experts about their background (position/ occupation, institution/ organization, nationality, age) and Madagascar work experience, i.e., what type of work related to the slash-and-burn problematic they have done, for how long they have worked in this field, whether they have worked with locals and if so, in which context. We checked these background variables to check whether they qualified as experts and to determine potential bias in ranking scores. For example, people which feel strongly attached to a place, younger people, as well as women are often more environmentally concerned than those with little place attachment, older people, and men, respectively [86]. Age and work experience were divided into categories (age: <25 years, between 25–35 years, 36–45 years, 46–65 years and > 65 years; work experience: <1 year, 1–5 years, 6–10 years, >10 years). Additionally, participants could provide any other background information about themselves they considered relevant to define their expertise. In part II, experts were introduced to the ranking scheme and then asked to rank the knowledge gaps by importance to pinpoint key knowledge gaps. The ranking categories reached from 0 (no priority) to 3 (extremely high priority). To accommodate for the lack of expert engagement in compiling the list of knowledge gaps, experts were given the opportunity to suggest additional knowledge gaps which they considered more important than those they ranked as a knowledge gap of category 3. A copy of the full questionnaire, including the cover letter and a last page with options for comments, optional email provisioning to receive information about the project outcome, thanks, and the declaration of consent can be found in the Questionnaire provided in S1.

Questionnaires were created in Adobe Acrobat professional. The ranking categories were fixed to the ranking scheme (0–3), text sections allowed unlimited words and no grammar correction was chosen to avoid false auto-correction. The survey was conducted in February and March 2019. To keep the identity of our experts anonymous, responses of individuals are not presented in a personalized way but only in non-assignable categories (Table S2a,b).

Results of this survey are of mainly descriptive nature. We ranked knowledge gaps by their mean (highest to lowest). Most background variables were unsuitable to use for statistical tests as we did not have enough scores for each category (position/occupation, institution/organization, nationality, age, years of work experience in western Madagascar) or because almost all participants belonged to just one category, i.e., 82% had >10 years of work experience and 96% had work experience with locals. The variables "position/occupation" and "institution/organization" had too many levels to be tested. Other variables with several levels were merged together in the following way: The five age groups were merged into young (>25 and 25–35), medium age (36–45) and old (46 –65, >65) age. The variable 'nationality' was reduced to two levels: Malagasy and other nationalities. Background variables with sufficiently even distribution were tested for their influence in ranking the knowledge gaps using Kruskal–Wallis tests with Holm's adjusted p-values controlling for family-wise Type I errors [87,88]. All analyses were performed in R 3.6.1 [89]. Due to well-known interlinkages of our focal SDGs (2, 10 and 15) to other SDGs, various ABTs and the 2 °C target, e.g., [10,12], we linked our top 10 knowledge gaps to other relevant targets of these agreements, thus pointing out which targets of these agreements will be better reached when filling these priority knowledge gaps.

3. Results

3.1. Background Information of Participants

A total of 51 participants returned the questionnaires (Table S2a), 67% of them were male. Most respondents were between 25–35 years old (37%), followed by those between 36–45 (29%) and 46–65 years (26%) which corresponded well to the young Malagasy society. Younger than 25 and older than 65 years were, in both age groups, only 4% of all participants. 77% of all participants were of Malagasy origin. The spectrum of positions covered various backgrounds. Most were researchers (25%), followed by students and managers (14% each), technical staff and coordinators (8%), lecturers (6% each), and others (<5% each). Accordingly, participants were associated with a range of organizations. Most were from universities (29%), followed by NGOs (26%), research centers (10%), and the UN (6%). All others contributed less than 5% each (e.g., governments, associations, etc.; see Table S2b for details). 82% had more than 10 years of work experience in western Madagascar while no participant had <1-year experience. Over 96% had work experience with locals.

3.2. Top 10 Knowledge Gaps across Categories

In two cases, knowledge gaps had equal mean scores, so that twelve knowledge gaps made it into the top 10 (Table 1). These consisted of four political, three ecological, three socio–economic and two general knowledge gaps (Table 1, Figure 2). With a mean of 2.68, more knowledge about "Strategies on how to improve justice/fairness/enforcement of laws/rules" was considered most important by a majority of participants. The knowledge gap with the second highest mean (2.67) was "Role of corruption in illegal activities (also beyond logging) and ways to reduce corruption". In the third position ranked the ecological knowledge gap "Appropriate forest restoration methods in conjunction with biodiversity protection and sustainable use" (mean 2.65). The fourth highest ranking knowledge gap was a socio–economic one: "Economic benefits for local small-holder farmers from biodiversity, e.g., potential of ecotourism, payment for ecosystem services (PES), and other off-set schemes on their well-being" (mean 2.61). The political knowledge gaps "Strategies on how to improve security from violence/theft/corruption" (mean 2.59) and "Strategies to improve long-term funding" (mean 2.50) ranked fifth and sixth. Rank seven was shared by an ecological ("Appropriate livestock management practices and fire regimes") and a socio–economic ("Effectiveness of education and awareness-raising on biodiversity conservation") knowledge gap (mean 2.45). The eight highest ranking knowledge gap was an ecological one: "Ecosystem services (ES) at risk from slash-and-burn as well as associated extractive activities" (mean 2.44). Two knowledge gaps shared rank nine (mean 2.43), a socio–economic ("Traditional knowledge about sustainable natural resource use") and a general one ("Frequent and regular scenario updates based on long-term monitoring"). The lowest ranking knowledge gap of the Top 10 list was a general one: "Interdisciplinary work to generate most comprehensive data sets" (mean 2.41). In Table S3 we provide the ranking of all 71 knowledge gaps.

Table 1. Linkages between the top 10 knowledge gaps ordered by mean ranking score and relevant Sustainable Development Goal (SDGs)[1], Achi Biodiversity Targets (ABTs)[2] and the 2 °C target[3] of the Paris Climate Agreement. Note that due to the same mean values of some knowledge gaps, a rank of the top 10 rankings can harbor more than one knowledge gap.

Top	Category	Knowledge Gap	Mean	SD	SDG	ABT	2 °C Target
1	Political	Strategies on how to improve justice/fairness/enforcement of laws/rules	2.68	0.583	10, 12, 16	4	
2	Political	Role of corruption in illegal activities (also beyond logging) and ways to reduce corruption	2.67	0.589	10, 12, 16	4	
3	Ecological	Appropriate forest restoration methods in conjunction with biodiversity protection and sustainable use	2.65	0.627	12, 13, 15	4, 7, 14, 15	yes
4	Socio-economic	Economic benefits for the local small-holder farmers from biodiversity: e.g., potential of ecotourism, PES and other offset schemes on their well-being	2.61	0.695	8, 10, 15	11, 14	
5	Political	Strategies on how to improve security from violence/theft/corruption	2.59	0.726	10, 16		
6	Political	Strategies to improve long-term funding	2.51	0.731		20	
7	Ecological	Appropriate livestock management practices and fire regimes	2.45	0.832	13, 15	7	yes yes
	Socio-economic	Effectiveness of education and awareness raising on biodiversity conservation	2.45	0.832	4	1	
8	Ecological	Ecosystem services (ES) at risk from slash-and-burn as well as associated extractive activities	2.44	0.760	13, 15	7	yes
9	General	Frequent and regular scenario updates based on long-term monitoring	2.43	0.755		7, 20	
	Socio-economic	Traditional knowledge about sustainable natural resource use	2.43	0.728		7, 18, 19	
10	General	Interdisciplinary work to generate more comprehensive data sets	2.41	0.669	17	18, 19, 20	

[1] SDG 1: No poverty, SDG 2: Zero hunger, SDG 3 Good health and well-being, SDG 4: Quality education, SDG 5 Gender equality, SDG 6 Clean water and sanitation, SDG 7: Affordable and clean energy, SDG 8: Decent work and economic growth, SDG 9: Industry, innovation and infrastructure, SDG 10: Reduced inequalities, SDG 11: Sustainable cities and communities, SDG 12: Responsible consumption and production, SDG 13: Climate action, SDG 14: Life below water, SDG 15: Life on land, SDG 16: Peace and justice, strong institutions, SDG 17: Partnerships for the goals. 2 ABT 1: Awareness of biodiversity increased, ABT 2: Biodiversity values integrated, ABT 3: Incentives reformed, ABT 4: Sustainable production and consumption, ABT 5: Habitat loss halved or reduced, ABT 6: Sustainable management of aquatic living resources, ABT 7: Sustainable agriculture, aquaculture and forestry, ABT 8: Pollution reduced, ABT 9: Invasive alien species prevented and controlled, ABT 10: Ecosystem vulnerable to climate change, ABT 11: Protected Areas, ABT 12: Reducing the risk of extinction, ABT 13: Safeguarding genetic diversity, ABT 14: Ecosystem services, ABT 15: Ecosystem restoration and resilience, ABT 16: Access to and sharing benefits from genetic resources, ABT 17: Biodiversity strategies and action plans, ABT 18: Traditional knowledge, ABT 19: Sharing information and knowledge, ABT 20: Mobilizing resources from all sources. 3 Until 2100 keep warming well below 2 °C.

3.3. Links of the Top 10 Knowledge Gaps to the SDGs, ABTs and 2 °C-Target of the Paris Climate Agreement

While we focused on SDG 2, 10 and 15 when identifying our knowledge gaps at the global scale, we are aware of the interlinkages of these SDGs with other SDGs, various ABTs and the 2 °C target [10,12]. Therefore, we linked the knowledge gaps identified for our focus area to other relevant SDGs, ABTs and the 2 °C target (Table 1, Figure 2). Ecological knowledge gaps identified within the top 10 knowledge gaps were related SDG 12 (Responsible consumption and production), SDG 13 (Climate change) and to SDG 15 (Life on Land), ABT 7 (Sustainable agriculture, aquaculture and forestry) and the 2 °C target (Table 1, Figure 2). Political knowledge gaps were most closely related to SDG 10 (Reduced inequalities), SDG 12 (Responsible consumption and production) and SDG 16 (Peace and justice, strong institutions), as well as ABT 4 (Sustainable production and consumption) and ABT 20 (Mobilizing resources from all sources) (Table 1, Figure 2). Socio–economic knowledge gaps were linked to SDG 4 (Quality education), SDG 8 (Decent work and economic growth), SDG 10 (Reduced inequalities) and SDG 15 (Life on land), as well as to ABT 1 (Awareness of biodiversity increased), ABT 11 (Protected areas), ABT 14 (Ecosystem services), ABT 18 (Traditional knowledge) and ABT 19 (Sharing information and knowledge) (Table 1, Figure 2). General knowledge gaps are linked to SDG 17 (Partnerships for the goals), ABT 7 (Sustainable agriculture, aquaculture and forestry), ABT 19 (Sharing information and knowledge) and ABT 20 (Mobilizing resources from all sources).

Figure 2. Overview of the expert-identified top 10 knowledge gaps (wording shortened for display) concerning the slash-and-burn problematic in the dry deciduous forests of Western Madagascar grouped by category (ECOL, POLITICAL, GENERAL, EOCEC) and linked to the SDGs (SDG symbols), Achi Biodiversity Targets (ABTs) (ABT symbols) and 2 °C target of the Paris Climate Agreement (COP 11 logo as symbol). ECOL: Ecological; SOCEC: Socio-economic. The full wording of the knowledge gaps can be seen in Table 1. The numbers of the thin tiles represent the ranking position within the top 10.

3.4. Effects of Gender, Nationality and Age on Ranking Knowledge Gaps

Most knowledge gaps were ranked similarly by male and female participants (Table S4a). This was true for all socio–economic and ecological, and all but one general ("Better data quality/reliability"), as well as two political knowledge gaps ("Strategies on how accountability of institutions/governments can be strengthened", "Effects of conservation activities and sustainable use of biodiversity on political achievements") which were ranked higher by women than men (Table S4a, Figure S1a). However, this affected none of the top 10 knowledge gaps (Table 1).

Rankings of knowledge gaps by Malagasy and other nationalities only differed for one general knowledge gap: "Frequent and regular scenario updates based on long-term monitoring") which was ranked higher by participants of Malagasy origin than those of other nationalities (Table S4b). We could not detect any differences in the ranking of knowledge gaps for different age groups (Table S4c, Figure S1c).

3.5. Additional Knowledge Gaps Suggested by Participants

For each knowledge gap category, the participants were given the option to list additional knowledge gaps which they considered of greater importance than those they gave the highest-ranking score. About one third of the participants used this option. Only one entirely new knowledge gap was listed as an ecological as well as a socio–economic knowledge gap by three participants: the impacts of climate change-driven human migration from southwestern to western Madagascar on socio–economic problems and its impacts on natural resources.

4. Discussion

The background information provided by our 51 experts demonstrates that they qualified as such given their backgrounds including work experience. The experts' backgrounds also shows that our sourced experts cover a wide range of positions and institutions so that bias towards interests of a particular group of researchers or decision makers is unlikely. Our top 10 list of key knowledge gaps addresses knowledge gaps from all three spheres of sustainability consisting of three ecological, three socio–economic knowledge gaps, four political, as well as two general knowledge gaps. None of the identified top 10 knowledge gaps was directly related to SDG 2 (Reduce Hunger), only four to SDG 10 (Reduce inequality) and three to SDG 15 (Life on Land). In the following paragraphs, we will first discuss the top 10 knowledge gaps by linking them to all SDGs, ABTs and the 2 °C target of the Paris Climate Agreement and then show recent local or larger scale attempts starting to address these knowledge gaps. Acknowledging the implications of our study and its limitations, we will provide suggestions for further work.

4.1. Top 10 Knowledge Gaps

The first two overall top priority knowledge gaps address the need to find "strategies to improve justice, fairness, enforcement of laws/rules" and to "gain more knowledge about the role of corruption as well as how to reduce it". Both knowledge gaps are strongly linked to another knowledge gap in the top 10 ranking: "Strategies to improve security from violence/theft/corruption" (rank 5). All three are addressed in SDG 16—one of the nine SDGs Madagascar failed to achieve according to the Africa SDG Index and Dashboard Report 2018 [45]. The issue of justice and fairness is also related to SDG 10 (Reduced inequalities), yet for African countries, no metrics exist for the achievement of this SDG [45]. So far, the issue of corruption has been completely lacking from the ABTs [15]. Lack of law enforcement, Zebu cattle theft and burglaries remain important issues for the rural population and the attempts to reduce corruption via an Anti-corruption Commission in 2002, Anti-Corruption Agency in 2004, or a governmental decree to ban illegal logging in 2005 have not helped to significantly reduce illegal activities [39,90]. Achieving better law/rule enforcement and fighting corruption is difficult as politicians, higher officials and local elites are often themselves involved in bribery or put no efforts in law enforcement for monetary, social or other benefits [91–94]. Corruption considerably undermines the protection of nature [95] as it reduces law enforcement and investments by international conservation agencies [91,95]. Since poor people often rely on the extraction of resources from protected areas to secure protein supply or to cover other subsistence needs, conservation laws and rules are often perceived as unfair and have been internationally criticized as "green-grabbing" [23,96,97]. This issue is addressed in the socio–economic knowledge gap "Economic benefits for local small-holder farmers from biodiversity" ranking 4th overall (see further below).

Internationally, the most well-known type of corruption in Madagascar is probably associated with the illegal trade of precious wood species (particularly *Dalbergia* spp.) due to the involvement of politicians and international media attention. Madagascar's new president Andry Rajoelina who was elected in December 2018 promised to make the fight against corruption a priority [98]. However, given his past involvement in illegal rosewood trade, his agreement to reinstate the ban of rosewood trade that lacked any reinforcement [41], and government corruption levels which spiked under his de facto presidency between 2009 and 2013, many, particularly the international community, doubt his promise to curb corruption [99].

Since the illegal rosewood trade occurs predominantly in eastern Madagascar [100] and also because it is only one of many corruption issues, we deliberately included "corruption beyond logging" in this knowledge gap. Corruption affects all sorts of sectors, for example funds destined for education [101], the church [102], undermining of local land rights, and access associated with agribusinesses and mining operations [39]. Political instability and multiple political crises have resulted in difficulties to establish or monitor new policies [54]. This has been accompanied by an increase in deforestation for agriculture [63] as well as in other types of illegal resource exploitation [33,42,69,100]. Those trying to fight corruption (e.g., local forest guards, environmental groups, members of watchdog organizations, researchers) live dangerous lives as they have to fear being evicted as "rebellions" or being confronted with death threats to them and their family members [96,103].

Corruption indicators used to evaluate the SDG 16 (e.g., number of victims of intentional homicide, conflict-related deaths per 100,000 population, proportion of children who experienced any physical punishment and/or psychological aggression) lack many of the above described components of corruption issues in Madagascar [104]. The International Corruption Perception Index relies on experts' opinions regarding transparency, accountability and corruption in the public sector [44], even though Transparency International experts are mainly business people who seem ill-suited to represent locally affected residents. Involving the local communities via questionnaires or interviews as done by Gore et al. (2013) for a different conservation area in Madagascar would give a better understanding of the "hotspots" of corruption activities and thus easier ways to tackle these issues right where they occur. More general strategies to fight corruption would be fair wages, stringent accounting procedures and management partnerships [95]. Media coverage of corruption activities and the work of brave activists to fight them may also be helpful [101], as is pressure from the international community [91].

The highest ranking ecological knowledge gap (third highest overall rank) directly addresses the need to know more about appropriate restoration beneficial to the protection and sustainable use of biodiversity. Restoration is key to providing essential ecosystem services (ABT 14) and has been given increased attention by the Global Partnership for Forest Restoration and the Bonn Challenge (e.g., Ockenden et al. 2018). Ecosystem restoration and resilience is explicitly mentioned in ABT 15, but not in the targets of SDG 15. It is linked to SDG 12 and ABTs 4 and 7 due its consequences for sustainable production in terms of most sustainable reforestation methods. Technological advances (Perring et al. 2015), green finance options (FAO and Global Mechanism of the UNCD, 2015) and compensation measures, particularly in regards to telecoupling effects, can all support restoration activities (e.g., Ockenden et al. 2018). Restoration projects require long-term efforts of monitoring as well as associated adjustments to effectively protect biodiversity and enable sustainable use of resources within restored areas. Given that some species will be unable to cope with even slight anthropogenic disturbance, some restoration areas, particularly corridors enabling migration between remaining habitat patches, will have to meet the needs of these sensitive species. Other restoration areas allowing for human resource utilization may still provide habitats for a number of less sensitive species. "Strategies to secure long-term funding" (rank 6) will be necessary to ensure that restoration will be sustainable.

If wisely done, reforestation can be extremely beneficial for biodiversity by providing suitable habitat connected by corridors or stepping stones [105,106]. Madagascar committed to the Bonn Challenge to reforest four million hectares of forest, but little progress has been made and the benefits only address the economy and climate [40,107]. Madagascar's National Biodiversity Strategy and Action Plan (NBSAP) lists restoration of at least 15% degraded habitats as one of its five strategic goals (Objective 15 in [108], p. 89). As a solution it mentions agro–ecological techniques as "effective tools for the degraded vegetation" but does not provide any details on what these tools to restore degraded vegetation are. Similarly, "appropriate strategies are set up to safeguard these ecosystems [...] for human well-being especially local communities through restoration activities" are mentioned, but actual strategies on how this can be achieved are only vaguely mentioned: "Tools or Handbook of conservation and/or Ecological restoration of various existing tropical forest types are developed" [ibid.] (p. 106) and "Number of recovery programs for protected areas of degraded ecosystems is developed and implemented" [ibid.] (p. 114). The knowledge gap concerning appropriate restoration strategies is also linked to SDG 13 and the 2 °C target of the Paris Climate Agreement as restoration will help mitigate climate change due to increased carbon storage, a decreased albedo and associated effects caused by revegetation measures [105,109,110].

The socio–economic knowledge gap considered most important (overall rank 4), "Economic benefits for local small-holder farmers from biodiversity", is directly related to SDG 8 (Decent work) in combination with the protection of biodiversity (SDG 15, ABT 11) and ecosystem services (ABT 14) and explicitly addressed as important in Madagascar's NBSAP [Objective 2 in 108] (p. 70f). Herein, it has been acknowledged that a fair distribution of benefits of ecosystem services requires more research, and where payment schemes for ecosystem services have been successful these approaches should be used as a guideline and adjusted to local contexts [65,111–113]. To date, most of these payments never covered the opportunity costs locals had to endure for the conservation of biodiversity, patrols were reluctant to convict fellow community members of illegal actions, and positive trends were reported although forest degradation actually increased [111,114,115].

Payments for ecosystem services like those within the REDD program (Reduced Emissions from Deforestation and Forest Degradation) can only be successful under certain conditions [115]. MacKinnon et al. (2017) reported that the amount of money reaching a community differed strongly between methods and that projects directly compensating for the loss of forest resources (e.g., by fish farming or bee keeping) were the most successful. Anticipated and promoted benefits for local people through tourism [69] rarely materialized [48,116], although there have been some cases of livelihood benefits from eco-tourism in a few tourism hotspots [114]. However, usually these benefited only a few people and not whole communities [102,117].

More knowledge regarding "strategies to improve security from violence/theft/corruption" has been the fifth highest ranking knowledge gap. This comes as no surprise, as Malagasy people have also ranked crime and insecurity as top priority issues for the government [118]. This knowledge gap concerns aspects of inequality (SDG 10) as poor people are most affected by these issues, but also to SDG 16 as strong institutions are necessary to tackle these issues. To avoid cattle theft, which is considered as "extremely worrisome" [119] cattle are kept hidden in forests, which promotes forest degradation and hampers forest regeneration [39]. People who have no choice but to continue to use forest resources for their survival also fear violence and fines from forest police [119]. Since law enforcement is largely lacking, laws are no deterrent for committing crimes in Madagascar, but an increase in law enforcement personnel locally and at least temporally decreases cattle theft (see [120] and references therein). Residents from villages with missions, police or military presence feel safer [119]. However, in remote areas, governmental control is particularly weak (or even completely lacking) [103]. While cattle theft is usually practiced by well-organized groups which cooperate with local authorities, burglaries and crop theft are rather associated with poverty and hunger of the rural poor as a coping strategy under high survival risk [120].

"Identification of strategies to improve long-term funding" has been identified as the 6th highest ranking knowledge gap and relates to all four spheres of sustainability [82]. Limited and short-term funding have been impeding long-term monitoring and successful participatory strategies [119]. Securing long-term funding for research, environmental as well as human aid projects is a global problem which is why this issue is picked up in ABT 20 (Mobilizing resources from all sources). It is strongly tied to the issue of short election cycles, awareness of the benefits of long-term funding instead of funding invested in many short-term projects and an issue of changes in staff when running long-term projects.

We combined the aspects of appropriate livestock management (here zebu cattle farming) and fire regimes in one knowledge gap (see Table 1, rank 7) as both have experienced little research and both are closely interrelated: grasslands, woodlands and forests are burnt to provide fresh pastures for zebu cattle and even where fire is not deliberately set, these pasture fires often escape unintentionally and burn adjacent habitats [39,121–123]. This knowledge gap is linked to the protection of biodiversity (SDG 15, ABT 7), sustainable management (SDG 12, ABT 4 and 7) and also to climate change (SDG 13, 2 °C target) as fire and cattle farming increase atmospheric CO_2-levels. Research needs for "appropriate management of livestock and fire regimes" are only partly addressed in Madagascar's National Biodiversity Strategy and Action Plan [108]: "programs aimed at strengthening the control of bush fires" and those that minimize fire impacts in areas with significant biodiversity by creating effective buffer zones around protected areas and more training of fire extinguishing personnel are explicitly mentioned [ibid.] (p. 107). Appropriate and more sustainable livestock management are not addressed though, indicating that this issue has not even gained due attention.

Zebu farming has a negligible role as a protein source, but cattle have important cultural value (slaughtered for special cultural occasions) and herd size indicates social status [39,122,124]. Zebu cattle are also used for agricultural labor and transport [125] and as an insurance asset to buffer income loss during times of hardship like droughts or low crop market prices [119,122,126]. The cultural importance of zebu cattle is particularly high for temporary Antandroy migrants from the south who use slash-and-burn agriculture for cash crops (particularly corn) to purchase cattle and thus increase their social status when returning home [119,125]. Climate change has already induced higher frequency and intensity of droughts in southern Madagascar which has led to increased migration of Antandroy from southern Madagascar into central western Madagascar severely increasing the pressure on the dry forests [39,122]. Traders involved in illegal corn and peanut plantation farms actively lure these migrants in for illegally clearing the forests for them by promising them quick and easy payment for each hectare of forest cleared as well as covering their travel and accommodation costs [127]. This climate change-induced migration issue has been the only additional knowledge gap suggested by several experts in our study. We agree that more knowledge about this issue and strategies to mitigate it via prosecution of and high fines for traffickers have a very high priority and that it should be on the priority list of current decision makers [40].

The third highest ranking ecological knowledge gap (overall rank 8) concerned ecosystem services at risk from slash-and-burn farming. Like the previous ecological knowledge gap concerning livestock and fire management, it is linked to biodiversity (SDG 15, ABT 7) and ecosystem service loss (ABT 14), as well as to consequences of fire on CO_2-levels and hence climate change (SDG 13, 2 °C target). We principally know that dry forests provide essential ecosystem services and that these are at risk from slash-and-burn farming but ranking this knowledge gap so high indicates that our understanding of which ecosystem services are affected with what potential cascading consequences is not yet perceived as being sufficient. Madagascar's NBSAP is strongly committed to ensure future provisioning of ecosystem services through the protection of 10% of terrestrial ecosystems, sustainable management (including certification), sustainable tourism, compensation activities like restoration, as well as additional studies to develop and implement appropriate strategies [108].

This knowledge gap has direct consequences for the sociological knowledge gap concerning the effectiveness of education and awareness-raising on biodiversity conservation linked to SDG 4

(Quality education) and ABT 1 (Awareness of biodiversity has increased). People who understand the implications of their actions are less likely to undertake them if they anticipate negative consequences for themselves or their families. Creating this understanding is particularly difficult when the effects of actions are complex and only visible in the more distant future. Better education may increase chances of income diversification [128] and improve community forest management [113]. Both, education and awareness raising are essential for the protection of biodiversity as they can lead to more understanding of ecosystem services nature provides [129]. For example, Reibelt, et al. [130] have shown that in south–western Madagascar, awareness of local people rises through direct contact with endangered species on which they rely for subsistence.

One of the two knowledge gaps at rank 9 concerns the improvement of sustainable methods by "more frequent and regular scenario-updates based on long-term monitoring". This general knowledge gap is of course directly related to the protection of biodiversity (SDG 15, ABT 7). Building the necessary trust for participatory approaches, negotiating rules and calculating long-term costs and benefits of biodiversity loss are all time-consuming activities. The efforts required for collecting solid data at realistic temporal scales need to be considered in biodiversity strategies and action plans (ABT 17). We considered this knowledge gap to be a general one as long-term monitoring for more frequent and regular scenario updates is also important in socio–economic and political planning, like long-term costs and successes of education programs (SDG 4) or the effectiveness of PES on political decisions.

Demonstrating how interrelated our identified key knowledge gaps are, the knowledge gap concerning appropriate strategies for restoration (overall rank 3) also relies on long-term monitoring—which again depends on long-term funding (rank 6), and consequently they all are strongly linked to ABT 20 ("Mobilizing resources from all sources").

The other knowledge gap at rank 9 "Traditional knowledge about sustainable natural resource use" belongs to the socio–economic sphere of sustainability and is directly linked to ABT 18 (Traditional knowledge): "In 2015, the initiatives put in place to protect traditional knowledge, innovations and practices of local communities relevant to biodiversity. The traditional sustainable use of biodiversity and their contribution to conservation are respected, preserved and maintained" which has been given funding priority by Madagascar's NBSAP when compared to restoration (Objective 15) and the evaluation of ecosystem services for PES schemes (Objective 2) (both in [108], (p. 154f)).

We need to become aware of traditional knowledge before it will be lost and to find out how traditional conservation values overlap with those of scientists [131]. Local people's knowledge has been shown to provide important insights for sustainable land management and use of resources in south-western Madagascar [132]. Several authors have shown that to achieve effective conservation in Madagascar, every ethnic group should be integrated from the start [133,134]. More use should be made of existing forms of traditional, local agreements and institutions, because such local rules are generally more adhered to [133,135].

The lowest ranking knowledge gap within our top 10 list "Interdisciplinary work to generate most comprehensive data sets" is strongly linked with the higher ranking ones. It is in some ways addressed in SDG 17 (Partnerships of the goals) and ABT 19 (Sharing information and knowledge) as well as ABT 20 (Mobilizing resources from all sources). In 2015, the journal Nature dedicated a special issue on the topic of interdisciplinarity demonstrating why "scientists must work together to save the world" [136]. However, interdisciplinarity should go beyond that of scientists and–in a transdisciplinary way–include knowledge from other sources like that of local/indigenous people or organizations with long-term on the ground experiences. Interdisciplinary work has been considered particularly important for studying the ecological impacts of climate change [137]. Finding ways to deal with environmental challenges associated with such impacts and their consequences for human well-being requires incorporating social and human-centered approaches by using participatory approaches [ibid.].

Most knowledge gaps which focused on generating more data within their own discipline received the lowest mean scores (Table S3), potentially indicating that our experts consider it more important to make use of existing data by using it in an interdisciplinary way than investing money into research of deficient, but highly specific research fields. A focus on improving decision-making and a call for more action rather than further investment in data collection has also been proposed by others [138]. However, we agree with Stuart et al. [139] that investing into more data generation, particularly those concerning threatened and data deficient species, remains important for conducting appropriate actions and adjusting management strategies as needed.

4.2. Effects of Different Expert Backgrounds on Ranking Scores

Our study showed that our top 10 list of key knowledge gaps would not have differed, would we have conducted this study without experts from other countries, only with women or only with experts of a certain age group, except for the knowledge gap concerning frequent and regular scenario updates which was considered of higher importance by Malagasy than other nationalities (Table S4, Figure S1b). Given the influence of place attachment, gender and age on the judgement of conservation related issues which has been pointed out by others [86], we are pleased about predominantly Malagasy experts ranking our knowledge gaps, about the relatively even gender ratio and that the bulk of our experts were medium and older aged and had long working experience in the area.

4.3. Limitations, Implications and Recommendations for Future Work

We are aware of the limitations of this study. To keep the questionnaire short, some knowledge gaps were already combined. Ideally, knowledge gaps should be kept separate and maybe in addition also presented in combination to give participants the option to score whether they found the combination more important than its parts. We tried to accommodate for this by giving the opportunity to express this in the comments sections or by listing additional highest priority knowledge gaps. Network sampling of experts as well as expert consultations in a relatively short time frame, which are typical for many projects, always have limitations as they are rarely comprehensive [84]. Our questionnaires were in English; future studies should include versions in French and potentially even Malagasy to be more inclusive. For other case studies, particularly those with more time and resources as well as those with a narrower, i.e., local focus, we recommend involving experts as well as local people right from the beginning, i.e., when the questions to be ranked are gathered [21]. To avoid translation problems when involving local people, highly skilled translators will be necessary. These need to be trained not only in translating western conservation ideas into local language and values, but also and most likely more importantly, the locals' values, beliefs and proverbs into concepts that can be integrated into conservation policy [116]. Even participatory approaches involving community members cannot claim to be the gold standard [140]. Since not every community member can be involved, participation is usually only possible for high ranking representatives of communities. These are usually elderly men so that a gender and age bias in selecting key knowledge gaps may become an issue. However, we agree that, given more time and resources, a Delphi-like process (i.e., at least two rounds of questionnaires, each accompanied by accumulation of responses and anonymous feedback to participants) should be used to prioritize knowledge gaps [21]. Ideally, representatives of local communities should be involved, particularly when attempting knowledge gap prioritization at a local level. We tried to accommodate for the lack of expert engagement from the start by providing an option to list additional knowledge gaps. We also acknowledge that the number of participants was relatively low and dominated by academics due to the networks of the authors. Despite those limitations, we hope that our work substantially contributes to pinpointing the most urgent knowledge gaps which need to be addressed to secure local livelihoods and better protect biodiversity in the dry forests of western Madagascar–and hence better achieve the SDGs, ABTs and the 2 °C target of the Paris Climate Agreement. Our priority list of knowledge gaps should be considered a first attempt of identifying most urgent research needs and help funding bodies to streamline investments at the regional level (i.e., the dry forests of western Madagascar). Like the global research questions identified for example by Oldekopp et al. (2016), the knowledge gaps identified for the slash-and-burn problematic in western

Madagascar can be used at a regional level as a starting point for research project designs, collaborations and debates between academics, practitioners, politicians and stakeholders.

5. Conclusions

In our study on the identification of priority knowledge gaps to better protect biodiversity and simultaneously secure local livelihoods, we focused on the loss of dry forests by subsistence agriculture in western Madagascar. However, we are aware that large international agribusinesses and the mining industry, as well as oil extraction practices, may locally present a bigger threat to this biome than slash-and-burn farming [39], particularly seeing how these industries has wiped forests from other countries like for example Ghana [141]. It is of great concern that Madagascar's president Andry Rajoelina's recently announced cabinet mainly consists of technicians and business leaders [98] and that he declared to follow Ghana in its recent advances in economic development [142]. Conservationists are highly concerned about his priority of short-term economic gains over long-term security of natural resources and ecosystem services [99,143,144]. Scientists including ourselves therefore call on the new president to keep to his promise aiming to curb corruption and to make Madagascar a model for conservation, while simultaneously urging the international community to continue its financial support of protecting biodiversity in Madagascar [40,143,145–147].

Not surprisingly given the interlinkages of the goals and targets of the SDGs, ABTs and 2 °C target of the Paris Climate Agreement [10,12,78], the results of our study demonstrate that even though we started with a focus on SDGs 2, 10 and 15, the priority knowledge gaps identified concern a wide range of other SDGs and address issues necessary to be solved to better reach several ABTs as well as the 2 °C target.

Many characteristics of the problems and challenges found in Madagascar are comparable to those in other biodiversity-rich but economically disadvantaged countries [33]. We hope that our idea of focusing the search for key knowledge gaps on a regional scale (i.e., the dry forests of western Madagascar) and using expert involvement via questionnaires can be used as a blue-print, inspiring and aligning most urgent research projects, streamlining research funds and resources to local needs.

Supplementary Materials: The following are available online at http://www.mdpi.com/2071-1050/11/20/5695/s1: Questionnaire S1, Table S2: Background information of participating experts, Table S3: Complete ranking of all knowledge gaps, Table S4: Knowledge gaps rankings compared between (a) female and male participants, (b) Malagasy participants and those of other nationality, and (c) different age groups, Table S5: Full list of participants' suggested additional knowledge gaps, Figure S1. Differences in mean ranking scores by (a) female and male participants, (b) participants of Malagasy origin and those of other nationalities, and (c) participants of young (<25, 25–35), medium (35–45) and older (45–65, >65) age.

Author Contributions: Conceptualization: A.S.K.F. and L.S., data curation: A.S.K.F.; Formal analysis: A.S.K.F.; funding acquisition: L.S.; investigation: L.S., A.S.K.F.; methodology: A.S.K.F. and L.S.; project administration: L.S.; resources: L.S.; supervision: L.S.; writing—original draft preparation: A.S.K.F., L.S.; writing—review and editing: A.S.K.F. and L.S.

Funding: This research was conducted within the project "Towards a future sustainable world where climate, biodiversity and human well-being are safeguarded" (sustainCBW)", funded by the Leibniz Association (grant number SAS-2017-PIK-LFV) and the Zoological Research Museum Alexander Koenig.

Acknowledgments: We thank the sustainCBW project partners Kirsten Thonicke and Diana Sietz (Potsdam Institute for Climate Impact Research), Katrin Vohland and Jens Jetzkowitz (Museum für Naturkunde Berlin), Ana Paula Aguiar and Odirilwe Selomane (Stockholm Resilience Centre), Carolin Sperk (Think Tank for Sustainability, Berlin) for valuable comments in the initial phase of this project and Wolfgang Wägele (Zoological Research Museum Alexander Koenig, Bonn) for logistic support. Particular thanks go to Peter Kappeler, Christian Roos and Matthias Markolf (German Primate Center, Göttingen) who helped pilot trialing the questionnaire. Matthias Markolf also helped us spreading the questionnaires through his networks. We would also like to thank all the participating experts in this study for their time and effort to fill in our questionnaire, and our three anonymous reviewers whose comments and suggestions significantly improved an earlier version of this manuscript.

Conflicts of Interest: The authors declare no conflict of interest.

References

1. UN General Assembly. Transforming our World: The 2030 Agenda for Sustainable Development. Available online: http://www.refworld.org/docid/57b6e3e44.htm (accessed on 27 June 2018).
2. Secretariat of the Convention on Biological Diversity. Strategic Plan for Biodiversity 2011–2020 and the Aichi Targets Living in Harmony with Nature. Available online: http://www.cbd.int/doc/strategic-plan/2011-2020/Aichi-Targets-EN.pdf (accessed on 27 June 2018).
3. United Nations. The Paris Agreement. Available online: https://unfccc.int/resource/docs/2015/cop21/eng/l09r01.pdf (accessed on 27 June 2018).
4. Naeem, S.; Chazdon, R.; Duffy, J.E.; Prager, C.; Worm, B. Biodiversity and human well-being: An essential link for sustainable development. *Proc. R. Soc. B Biol. Sci.* **2016**, *283*. [CrossRef] [PubMed]
5. Bennett, E.M.; Cramer, W.; Begossi, A.; Cundill, G.; Díaz, S.; Egoh, B.N.; Geijzendorffer, I.R.; Krug, C.B.; Lavorel, S.; Lazos, E.; et al. Linking biodiversity, ecosystem services, and human well-being: Three challenges for designing research for sustainability. *Curr. Opin. Environ. Sustain.* **2015**, *14*, 76–85. [CrossRef]
6. Chen, Y.; Peng, S. Evidence and mapping of extinction debts for global forest-dwelling reptiles, amphibians and mammals. *Sci. Rep.* **2017**, *7*, 44305. [CrossRef] [PubMed]
7. Butchart, S.H.M.; Walpole, M.; Collen, B.; van Strien, A.; Scharlemann, J.P.W.; Almond, R.E.A.; Baillie, J.E.M.; Bomhard, B.; Brown, C.; Bruno, J.; et al. Global Biodiversity: Indicators of Recent Declines. *Science* **2010**, *328*, 1164–1168. [CrossRef]
8. Ceballos, G.; Ehrlich, P.R.; Barnosky, A.D.; García, A.; Pringle, R.M.; Palmer, T.M. Accelerated modern human–induced species losses: Entering the sixth mass extinction. *Sci. Adv.* **2015**, *1*. [CrossRef]
9. TEEB. *TEEB for Agriculture & Food: An Interim Report*; United Nations Environment Programme: Geneva, Switzerland, 2015.
10. Le Blanc, D. Towards integration at last? The sustainable development goals as a network of targets. *Sustain. Dev.* **2015**, *23*, 176–187. [CrossRef]
11. Nilsson, M.; Griggs, D.; Visbeck, M. Policy: Map the interactions between Sustainable Development Goals. *Nat. News* **2016**, *534*, 320. [CrossRef]
12. Schultz, M.; Tyrrell, T.; Ebenhard, T. *The 2030 Agenda and Ecosystems—A Discussion Paper on the Links between Aichi Biodiversity Targets and the Sustainable Development Goals*; SwedBio at Stockholm Resilience Centre: Stockholm, Sweden, 2016.
13. Spangenberg, J.H. Hot air or comprehensive progress? A critical assessment of the SDGs. *Sustain. Dev.* **2017**, *25*, 311–321. [CrossRef]
14. Stafford-Smith, M.; Griggs, D.; Gaffney, O.; Ullah, F.; Reyers, B.; Kanie, N.; Stigson, B.; Shrivastava, P.; Leach, M.; O'Connell, D. Integration: The key to implementing the Sustainable Development Goals. *Sustain. Sci.* **2017**, *12*, 911–919. [CrossRef]
15. Driscoll, D.A.; Bland, L.M.; Bryan, B.A.; Newsome, T.M.; Nicholson, E.; Ritchie, E.G.; Doherty, T.S. A biodiversity-crisis hierarchy to evaluate and refine conservation indicators. *Nat. Ecol. Evol.* **2018**, *2*, 775–781. [CrossRef]
16. Di Marco, M.; Chapman, S.; Althor, G.; Kearney, S.; Besancon, C.; Butt, N.; Maina, J.M.; Possingham, H.P.; Rogalla von Bieberstein, K.; Venter, O.; et al. Changing trends and persisting biases in three decades of conservation science. *Glob. Ecol. Conserv.* **2017**, *10*, 32–42. [CrossRef]
17. Global Policy Watch. Desperately Seeking Indicators. Available online: https://www.globalpolicywatch.org/blog/2018/11/01/desperately-seeking-indicators/ (accessed on 13 April 2019).
18. Schmidt-Traub, G.; Kroll, C.; Teksoz, K.; Durand-Delacre, D.; Sachs, J.D. National baselines for the Sustainable Development Goals assessed in the SDG Index and Dashboards. *Nat. Geosci.* **2017**, *10*, 547. [CrossRef]
19. Joppa, L.N.; O'Connor, B.; Visconti, P.; Smith, C.; Geldmann, J.; Hoffmann, M.; Watson, J.E.M.; Butchart, S.H.M.; Virah-Sawmy, M.; Halpern, B.S.; et al. Filling in biodiversity threat gaps. *Science* **2016**, *352*, 416–418. [CrossRef]
20. Shackleton, R.T.; Angelstam, P.; van der Waal, B.; Elbakidze, M. Progress made in managing and valuing ecosystem services: A horizon scan of gaps in research, management and governance. *Ecosyst. Serv.* **2017**, *27*, 232–241. [CrossRef]
21. Sutherland, W.J.; Adams, W.M.; Aronson, R.B.; Aveling, R.; Blackburn, T.M.; Broad, S.; Ceballos, G.; Côte, I.M.; Cowling, R.M.; Da Fonseca, G.A.B.; et al. One Hundred Questions of Importance to the Conservation of Global Biological Diversity. *Conserv. Biol.* **2009**, *23*, 557–567. [CrossRef]

22. Convention on Biological Diversity. United Nations Decade on Biodiversity. Available online: www.cbd.int/2011-2020/ (accessed on 30 April 2019).
23. Fairhead, J.; Leach, M.; Scoones, I. Green Grabbing: A new appropriation of nature? *J. Peasant Stud.* **2012**, *39*, 237–261. [CrossRef]
24. Oldekop, J.A.; Holmes, G.; Harris, W.E.; Evans, K.L. A global assessment of the social and conservation outcomes of protected areas. *Conserv. Biol.* **2016**, *30*, 133–141. [CrossRef]
25. Dudley, N.; Jonas, H.; Nelson, F.; Parrish, J.; Pyhälä, A.; Stolton, S.; Watson, J.E.M. The essential role of other effective area-based conservation measures in achieving big bold conservation targets. *Glob. Ecol. Conserv.* **2018**, *15*, e00424. [CrossRef]
26. Adams, W.M.; Aveling, R.; Brockington, D.; Dickson, B.; Elliott, J.; Hutton, J.; Roe, D.; Vira, B.; Wolmer, W. Biodiversity Conservation and the Eradication of Poverty. *Science* **2004**, *306*, 1146–1149. [CrossRef]
27. Andam, K.S.; Ferraro, P.J.; Sims, K.R.E.; Healy, A.; Holland, M.B. Protected areas reduced poverty in Costa Rica and Thailand. *Proc. Natl. Acad. Sci. USA* **2010**, *107*, 9996–10001. [CrossRef]
28. Kashwan, P. Inequality, democracy, and the environment: A cross-national analysis. *Ecol. Econ.* **2017**, *131*, 139–151. [CrossRef]
29. Holland, T.G.; Peterson, G.D.; Gonzalez, A. A cross-national analysis of how economic inequality predicts biodiversity loss. *Conserv. Biol.* **2009**, *23*, 1304–1313. [CrossRef] [PubMed]
30. Hicks, C.C.; Levine, A.; Agrawal, A.; Basurto, X.; Breslow, S.J.; Carothers, C.; Charnley, S.; Coulthard, S.; Dolsak, N.; Donatuto, J.; et al. Engage key social concepts for sustainability. *Science* **2016**, *352*, 38–40. [CrossRef] [PubMed]
31. Biedenweg, K.; Gross-Camp, N.D. A brave new world: Integrating well-being and conservation. *Ecol. Soc.* **2018**, *23*. [CrossRef]
32. Tabor, K.; Hewson, J.; Tien, H.; González-Roglich, M.; Hole, D.; Williams, J. Tropical Protected Areas Under Increasing Threats from Climate Change and Deforestation. *Land* **2018**, *7*, 90. [CrossRef]
33. Waeber, P.O.; Wilmé, L.; Mercier, J.-R.; Camara, C.; Lowry, P.P., II. How Effective Have Thirty Years of Internationally Driven Conservation and Development Efforts Been in Madagascar? *PLoS ONE* **2016**, *11*, e0161115. [CrossRef]
34. World Bank. *Madagascar: Poverty Assessment*; World Bank: Washington, DC, USA, 1996; Available online: http://documents.worldbank.org/curated/en/docsearch/report/14044 (accessed on 12 September 2019).
35. Selomane, O.; Reyers, B.; Biggs, R.; Tallis, H.; Polasky, S. Towards integrated social–ecological sustainability indicators: Exploring the contribution and gaps in existing global data. *Ecol. Econ.* **2015**, *118*, 140–146. [CrossRef]
36. Ploch, L.; Cook, N. *Madagascar's Political Crisis*; U.S. Congressional Research Service: Washington, DC, USA, 2012; Available online: http://www.fas.org/sgp/crs/row/R40448.pdf (accessed on 18 November 2018).
37. Hänke, H.; Barkmann, J.; Coral, C.; Kaustky, E.E.; Marggraf, R. Social-ecological traps hinder rural development in southwestern Madagascar. *Ecol. Soc.* **2017**, *22*. [CrossRef]
38. World Bank. *ID4D Country Diagnostic: Madagascar*; World Bank: Washington, DC, USA, 2017; Available online: http://pubdocs.worldbank.org/en/809191510763351833/Madagascar-ID4D-IMSA.pdf (accessed on 18 November 2018).
39. Waeber, P.O.; Wilmé, L.; Ramamonjisoa, B.; Garcia, C.; Rakotomalala, D.; Rabemananjara, Z.H.; Kull, C.A.; Ganzhorn, J.U.; Sorg, J.P. Dry forests in Madagascar: Neglected and under pressure. *Int. For. Rev.* **2015**, *17*, 127–148. [CrossRef]
40. Jones, J.P.G.; Ratsimbazafy, J.; Ratsifandrihamanana, A.N.; Watson, J.E.M.; Andrianandrasana, H.T.; Cabeza, M.; Cinner, J.E.; Goodman, S.M.; Hawkins, F.; Mittermeier, R.A.; et al. Last chance for Madagascar's biodiversity. *Nat. Sustain.* **2019**, *2*, 350–352. [CrossRef]
41. Innes, J.L. Madagascar rosewood, illegal logging and the tropical timber trade. *Madag. Conserv. Dev.* **2010**, *5*, 6–10. [CrossRef]
42. Randriamalala, H.; Liu, Z. Rosewood of Madagascar: Between democracy and conservation. *Madag. Conserv. Dev.* **2010**, *5*, 12. [CrossRef]
43. Institute for Economics & Peace. *Global Peace Index 2018: Measuring Peace in a Complex World*; Sydney, Australia, 2018. Available online: http://visionofhumanity.org/reports (accessed on 7 November 2018).
44. Transparency International. Corruption Perceptions Index 2017. Available online: https://www.transparency.org/news/feature/corruption_perceptions_index_2017 (accessed on 13 April 2019).

45. The Sustainable Development Goals Center for Africa and Sustainable Development Solutions Network. *Africa SDG Index and Dashboards Report 2018*; The Sustainable Development Goals Center for Africa and Sustainable Development Solutions Network: Kigali, Rwanda; New York, NY, USA, 2018.
46. Rasolofoson, R.A.; Nielsen, M.R.; Jones, J.P.G. The potential of the Global Person Generated Index for evaluating the perceived impacts of conservation interventions on subjective well-being. *World Dev.* **2018**, *105*, 107–118. [CrossRef]
47. Gardner, C.J.; Nicoll, M.E.; Birkinshaw, C.; Harris, A.; Lewis, R.E.; Rakotomalala, D.; Ratsifandrihamanana, A.N. The rapid expansion of Madagascar's protected area system. *Biol. Conserv.* **2018**, *220*, 29–36. [CrossRef]
48. Scales, I.R. The future of biodiversity conservation and environmental management in Madagascar: Lessons from the past and challenges ahead. In *Conservation and Environmental Management in Madagascar*; Routledge: London, UK, 2014; pp. 342–360.
49. Myers, N.; Mittermeier, R.A.; Mittermeier, C.G.; da Fonseca, G.A.B.; Kent, J. Biodiversity hotspots for conservation priorities. *Nature* **2000**, *403*, 853. [CrossRef] [PubMed]
50. Allnutt, T.F.; Asner, G.P.; Golden, C.D.; Powell, G.V.N. Mapping Recent Deforestation and Forest Disturbance in Northeastern Madagascar. *Trop. Conserv. Sci.* **2013**, *6*, 1–15. [CrossRef]
51. Razafimanahaka, J.H.; Jenkins, R.K.B.; Andriafidison, D.; Randrianandrianina, F.; Rakotomboavonjy, V.; Keane, A.; Jones, J.P.G. Novel approach for quantifying illegal bushmeat consumption reveals high consumption of protected species in Madagascar. *Oryx* **2012**, *46*, 584–592. [CrossRef]
52. Hockley, N.; Mandimbiniaina, R.; Rakotonarivo, O.S. Fair and equitable conservation: Do we really want it, and if so, do we know how to achieve it? *Madag. Conserv. Dev.* **2018**, *13*, 3–5. [CrossRef]
53. Scales, I.R. Farming at the Forest Frontier: Land Use and Landscape Change in Western Madagascar, 1896–2005. *Environ. Hist.* **2011**, *17*, 499–524. [CrossRef]
54. Stoudmann, N.; Waeber, P.O.; Randriamalala, I.H.; Garcia, C. Perception of change: Narratives and strategies of farmers in Madagascar. *J. Rural Stud.* **2017**, *56*, 76–86. [CrossRef]
55. Jarosz, L. Defining and Explaining Tropical Deforestation: Shifting Cultivation and Population Growth in Colonial Madagascar (1896–1940). *Econ. Geogr.* **1993**, *69*, 366–379. [CrossRef] [PubMed]
56. Ganzhorn, J.U.; Sorg, J.-P. *Ecology and Economy of a Tropical Dry Forest in Madagascar*; Deutsches Primatenzentrum: Göttingen, Germany, 1996.
57. Scales, I.R. The drivers of deforestation and the complexity of land use in Madagascar. In *Conservation and Environmental Management in Madagascar*; Routledge: London, UK, 2014; pp. 129–150.
58. Raharimalala, O.; Buttler, A.; Dirac Ramohavelo, C.; Razanaka, S.; Sorg, J.-P.; Gobat, J.-M. Soil–vegetation patterns in secondary slash and burn successions in Central Menabe, Madagascar. *Agric. Ecosyst. Environ.* **2010**, *139*, 150–158. [CrossRef]
59. Gay-des-Combes, J.M.; Robroek, B.J.M.; Hervé, D.; Guillaume, T.; Pistocchi, C.; Mills, R.T.E.; Buttler, A. Slash-and-burn agriculture and tropical cyclone activity in Madagascar: Implication for soil fertility dynamics and corn performance. *Agric. Ecosyst. Environ.* **2017**, *239*, 207–218. [CrossRef]
60. Ganzhorn, J.U.; Lowry, P.P.; Schatz, G.E.; Sommer, S. The biodiversity of Madagascar: One of the world's hottest hotspots on its way out. *Oryx* **2001**, *35*, 346–348. [CrossRef]
61. Reed, D.H. Impact of Climate Change on Biodiversity. In *Handbook of Climate Change Mitigation and Adaptation*; Chen, W.-Y., Suzuki, T., Lackner, M., Eds.; Springer: Berlin, Germany, 2017; pp. 595–620.
62. Hannah, L.; Dave, R.; Lowry, P.P.; Andelman, S.; Andrianarisata, M.; Andriamaro, L.; Cameron, A.; Hijmans, R.; Kremen, C.; MacKinnon, J.; et al. Climate change adaptation for conservation in Madagascar. *Biol. Lett.* **2008**, *4*, 590–594. [CrossRef]
63. Zinner, D.; Wygoda, C.; Razafimanantsoa, L.; Rasoloarison, R.; Andrianandrasana, H.T.; Ganzhorn, J.U.; Torkler, F. Analysis of deforestation patterns in the central Menabe, Madagascar, between 1973 and 2010. *Reg. Environ. Chang.* **2014**, *14*, 157–166. [CrossRef]
64. Mittermeier, R.A.; Myers, N.; Thomsen, J.B.; Da Fonseca, G.A.B.; Olivieri, S. Biodiversity Hotspots and Major Tropical Wilderness Areas: Approaches to setting conservation priorities. *Conserv. Biol.* **1998**, *12*, 516–520. [CrossRef]
65. Sommerville, M.; Jones, J.P.; Rahajaharison, M.; Milner-Gulland, E. The role of fairness and benefit distribution in community-based Payment for Environmental Services interventions: A case study from Menabe, Madagascar. *Ecol. Econ.* **2010**, *69*, 1262–1271. [CrossRef]

66. Schäffler, L.; Kappeler, P.M. Distribution and Abundance of the World's Smallest Primate, Microcebus berthae, in Central Western Madagascar. *Int. J. Primatol.* **2014**, *35*, 557–572. [CrossRef]
67. Ganzhorn, J.U.; Sommer, S.; Abraham, J.P.; Ade, M.; Raharivololona, B.M.; Rakotovao, E.R.; Rakotondrasoa, C.; Randriamarosoa, R. Mammals of the Kirindy Forest with special emphasis on *Hypogeomys antimena* and the effects of logging on the small mammal fauna. In *Ecology and Economy of a Tropical Dry Forest in Madagascar. Primate Report, 46–1*; Ganzhorn, J.U., Sorg, J.-P., Eds.; German Primate Center: Göttingen, Germany, 1996; pp. 215–232.
68. Bloxam, Q.M.C.; Behler, J.L.; Rakotovao, E.R.; Randriamahazo, H.J.A.R.; Hayes, K.T.; Tonge, S.J.; Ganzhom, J.U. Effects of logging on the reptile fauna of the Kirindy Forest with special emphasis on the flat-tailed tortoise (*Pyxis planicauda*). In *Ecology and Economy of a Tropical Dry Forest in Madagascar. Primate Report, 46–1*; Ganzhorn, J.U., Sorg, J.-P., Eds.; German Primate Center: Göttingen, Germany, 1996.
69. Schwitzer, C.; Mittermeier, R.A.; Johnson, S.E.; Donati, G.; Irwin, M.; Peacock, H.; Ratsimbazafy, J.; Razafindramanana, J.; Louis, E.E.; Chikhi, L.; et al. Averting Lemur Extinctions amid Madagascar's Political Crisis. *Science* **2014**, *343*, 842–843. [CrossRef]
70. Schäffler, L.; Saborowski, J.; Kappeler, P.M. Agent-mediated spatial storage effect in heterogeneous habitat stabilizes competitive mouse lemur coexistence in Menabe Central, Western Madagascar. *BMC Ecol.* **2015**, *15*, 7. [CrossRef] [PubMed]
71. Ganzhorn, J.U.; Fietz, J.; Rakotovao, E.; Schwab, D.; Zinner, D. Lemurs and the regeneration of dry deciduous forest in Madagascar. *Conserv. Biol.* **1999**, *13*, 794–804. [CrossRef]
72. Olson, D.M.; Dinerstein, E. The Global 200: Priority ecoregions for global conservation. *Ann. Mo. Bot. Garden* **2002**, *89*, 199–224. [CrossRef]
73. Langhammer, P.F.; Bakarr, M.I.; Bennun, L.; Brooks, T.M. *Identification and Gap Analysis of Key Biodiversity Areas: Targets for Comprehensive Protected Area Systems*; IUCN: Gland, Switzerland, 2007.
74. Rogers, H.M.; Glew, L.; Honzák, M.; Hudson, M.D. Prioritizing key biodiversity areas in Madagascar by including data on human pressure and ecosystem services. *Landsc. Urban Plan.* **2010**, *96*, 48–56. [CrossRef]
75. Corson, C. A history of conservation politics in Madagascar. *Madag. Conserv. Dev.* **2017**, *12*, 49–60. [CrossRef]
76. Sutherland, W.J.; Armstrong-Brown, S.; Armsworth, P.R.; Breeton, T.; Brickland, J.; Campbell, C.D.; Chamberlain, D.E.; Cooke, A.I.; Dulvy, N.K.; Dusic, N.R.; et al. The identification of 100 ecological questions of high policy relevance in the UK. *J. Appl. Ecol.* **2006**, *43*, 617–627. [CrossRef]
77. Oldekop, J.A.; Fontana, L.B.; Grugel, J.; Roughton, N.; Adu-Ampong, E.A.; Bird, G.K.; Dorgan, A.; Vera Espinoza, M.A.; Wallin, S.; Hammett, D.; et al. 100 key research questions for the post-2015 development agenda. *Dev. Policy Rev.* **2016**, *34*, 55–82. [CrossRef]
78. Pradhan, P.; Costa, L.; Rybski, D.; Lucht, W., Kropp, J.P. A Systematic Study of Sustainable Development Goal (SDG) Interactions. *Earth's Future* **2017**, *5*, 1169–1179. [CrossRef]
79. Wohlin, C. Guidelines for snowballing in systematic literature studies and a replication in software engineering. In Proceedings of the 18th International Conference on Evaluation and Assessment in Software Engineering, London, UK, 13–14 May 2014; pp. 1–10.
80. Gentles, S.J.; Charles, C.; Nicholas, D.B.; Ploeg, J.; McKibbon, K.A. Reviewing the research methods literature: Principles and strategies illustrated by a systematic overview of sampling in qualitative research. *Syst. Rev.* **2016**, *5*, 172. [CrossRef]
81. Lecy, J.; Beatty, K. Representative Literature Reviews Using Constrained Snowball Sampling and Citation Network Analysis. Available online: https://ssrn.com/abstract=1992601 (accessed on 16 July 2018).
82. Maxim, L.; Spangenberg, J.H.; O'Connor, M. An analysis of risks for biodiversity under the DPSIR framework. *Ecol. Econ.* **2009**, *69*, 12–23. [CrossRef]
83. Dick, M.; Rous, A.M.; Nguyen, V.M.; Cooke, S.J. Necessary but challenging: Multiple disciplinary approaches to solving conservation problems. *Facets* **2016**, *1*, 67–82. [CrossRef]
84. McLafferty, S.L. Conducting questionnaire surveys. In *Key Methods in Geography*; Clifford, N., French, S., Valentine, G., Eds.; Sage Publications Ltd.: Los Angelos, CA, USA, 2010; pp. 77–88.
85. Clifford, N.; French, S.; Valentine, G. (Eds.) *Key Methods in Geography*, 2nd ed.; Sage Publications Ltd.: Los Angelos, CA, USA, 2010; 545p.
86. Gifford, R.; Nilsson, A. Personal and social factors that influence pro-environmental concern and behaviour: A review. *Int. J. Psychol.* **2014**, *49*, 141–157. [CrossRef] [PubMed]

87. Quinn, G.P.; Keough, M.J. *Experimental Design and Data Analysis for Biologists*; Cambridge University Press: Cambridge, UK, 2002.
88. Holm, S. A Simple Sequentially Rejective Multiple Test Procedure. *Scand. J. Stat.* **1979**, *6*, 65–70.
89. R Core Team. R: A Language and Environment for Statistical Computing. Available online: https://www.R-project.org/ (accessed on 20 September 2019).
90. Francken, N.; Minten, B. Security and perceptions on justice in rural Madagascar. *Dynamics in Social Service Delivery and the Rural Economy of Madagascar: Descriptive Results of the 2004 Commune Survey*. Ilo Policy Report. 2005. Chapter 5. Available online: http://www.ilo.cornell.edu/images/chapter5.pdf (accessed on 20 September 2019).
91. Laurance, W.F. The perils of payoff: Corruption as a threat to global biodiversity. *Trends Ecol. Evol.* **2004**, *19*, 399–401. [CrossRef] [PubMed]
92. Francken, N.; Minten, B.; Swinnen, J.F. *Listen to the Radio! Media and Corruption: Evidence from Madagascar*; PRG Working Papers 31872; Katholieke Universiteit Leuven, LICOS – Centre for Institutions and Economic Performance: Leuven, Belgium, 2005.
93. Burnod, P.; Andrianirina Ratsialonana, R.; Teyssier, A. Processus d'acquisition foncière à grande échelle à Madagascar: Quelles régulations sur le terrain ? *Cahiers Agric.* **2013**, *22*, 33–38. [CrossRef]
94. Burnod, P.; Gingembre, M.; Andrianirina Ratsialonana, R. Competition over Authority and Access: International Land Deals in Madagascar. *Dev. Chang.* **2013**, *44*, 357–379. [CrossRef]
95. Garnett, S.T.; Joseph, L.N.; Watson, J.E.M.; Zander, K.K. Investing in Threatened Species Conservation: Does Corruption Outweigh Purchasing Power? *PLoS ONE* **2011**, *6*, e22749. [CrossRef]
96. Gore, M.L.; Ratsimbazafy, J.; Lute, M.L. Rethinking Corruption in Conservation Crime: Insights from Madagascar. *Conserv. Lett.* **2013**, *6*, 430–438. [CrossRef]
97. Pimm, S.L.; Ayres, M.; Balmford, A.; Branch, G.; Brandon, K.; Brooks, T.; Bustamante, R.; Costanza, R.; Cowling, R.; Curran, L.M.; et al. Can we defy nature's end? *Science* **2001**, *293*, 2207–2208. [CrossRef]
98. Anabel, Y. Andry Rajoelina Unveils Madagascar's New Cabinet. Available online: http://afrika-news.com/andry-rajoelina-unveils-madagascars-new-cabinet/ (accessed on 13 February 2019).
99. Laurance, W. Madagascar Elections: An Environmental Disaster in Waiting. Available online: https://www.aljazeera.com/indepth/opinion/madagascar-elections-environmental-disaster-waiting-181217095423818.html?fbclid=IwAR2Gj4TD1fa5xCn8yfD93W8XVsfNMOFEW5l0bqVwtYiAJf_M02LuK2h21g8 (accessed on 17 October 2018).
100. Barrett, M.A.; Brown, J.L.; Morikawa, M.K.; Labat, J.-N.; Yoder, A.D. CITES Designation for Endangered Rosewood in Madagascar. *Science* **2010**, *328*, 1109–1110. [CrossRef]
101. Francken, N.; Minten, B.; Swinnen, J.F.M. Media, Monitoring, and Capture of Public Funds: Evidence from Madagascar. *World Dev.* **2009**, *37*, 242–255. [CrossRef]
102. Freudenberger, K. *Paradise Lost? Lessons Learned from 25 Years of USAID Environment Programs in Madagascar*; Final Report U.S. Agency for International Development (USAID); International Resources Group: Washington, DC, USA, 2010.
103. Rakotomanana, H.; Jenkins, R.K.; Ratsimbazafy, J. Conservation challenges for Madagascar in the next decade. In *Conservation Biology: Voices from the Tropics*; Sodhi, S., Gibson, L., Eds.; John Wiley & Sons, Ltd: Oxford, UK, 2013; pp. 33–39.
104. UN. Global Indicator Framework for the Sustainable Development Goalsand Targets of the 2030 Agenda for Sustainable Development. Available online: https://unstats.un.org/sdgs/indicators/Global%20Indicator%20Framework%20after%20refinement_Eng.pdf (accessed on 15 April 2019).
105. Locatelli, B.; Catterall, C.P.; Imbach, P.; Kumar, C.; Lasco, R.; Marín-Spiotta, E.; Mercer, B.; Powers, J.S.; Schwartz, N.; Uriarte, M. Tropical reforestation and climate change: Beyond carbon. *Restor. Ecol.* **2015**, *23*, 337–343. [CrossRef]
106. Shoo, L.P.; Storlie, C.; Vanderlwal, J.; Little, J.; Williams, S.E. Targeted protection and restoration to conserve tropical biodiversity in a warming world. *Glob. Chang. Biol.* **2011**, *17*, 186–193. [CrossRef]
107. IUCN. The Bonn Challenge. Available online: http://www.bonnchallenge.org/content/madagascar (accessed on 23 July 2019).

108. Rabarison, H.; Randrianmahaleo, S.I.; Andriambelo, F.M.; Randrianasolo, H.L.; National Biodiversity Strategy and Action Plans 2015–2025. Decree N° 2016-128 of 2016, February 23 Adopting the National Biodiversity Strategy and Action Plans for Madagascar 2015 to 2025. Available online: https://www.cbd.int/doc/world/mg/mg-nbsap-v2-en.pdf (accessed on 21 November 2018).
109. Chen, I.-C.; Hill, J.K.; Ohlemüller, R.; Roy, D.B.; Thomas, C.D. Rapid range shifts of species associated with high levels of climate warming. *Science* **2011**, *333*, 1024–1026. [CrossRef] [PubMed]
110. Bastin, J.-F.; Finegold, Y.; Garcia, C.; Mollicone, D.; Rezende, M.; Routh, D.; Zohner, C.M.; Crowther, T.W. The global tree restoration potential. *Science* **2019**, *365*, 76–79. [CrossRef]
111. Sommerville, M.; Milner-Gulland, E.J.; Rahajaharison, M.; Jones, J.P.G. Impact of a Community-Based Payment for Environmental Services Intervention on Forest Use in Menabe, Madagascar. *Conserv. Biol.* **2010**, *24*, 1488–1498. [CrossRef]
112. Brimont, L.; Ezzine-de-Blas, D.; Karsenty, A. The cost of making compensation payments to local forest populations in a REDD+ pilot project in Madagascar. *Madag. Conserv. Dev.* **2017**, *12*, 25–33. [CrossRef]
113. Rasolofoson, R.A.; Ferraro, P.J.; Ruta, G.; Rasamoelina, M.S.; Randriankolona, P.L.; Larsen, H.O.; Jones, J.P.G. Impacts of Community Forest Management on Human Economic Well-Being across Madagascar. *Conserv. Lett.* **2017**, *10*, 346–353. [CrossRef]
114. Neudert, R.; Ganzhorn, J.U.; WÄTzold, F. Global benefits and local costs—The dilemma of tropical forest conservation: A review of the situation in Madagascar. *Environ. Conserv.* **2016**, *44*, 82–96. [CrossRef]
115. Danielsen, F.; Skutsch, M.; Burgess, N.D.; Jensen, P.M.; Andrianandrasana, H.; Karky, B.; Lewis, R.; Lovett, J.C.; Massao, J.; Ngaga, Y.; et al. At the heart of REDD+: A role for local people in monitoring forests? *Conserv. Lett.* **2011**, *4*, 158–167. [CrossRef]
116. Scales, I.R. Lost in translation: Conflicting views of deforestation, land use and identity in western Madagascar. *Geogr. J.* **2012**, *178*, 67–79. [CrossRef] [PubMed]
117. Scales, I.R. The future of conservation and development in Madagascar: Time for a new paradigm? *Madag. Conserv. Dev.* **2014**, *9*, 5–12. [CrossRef]
118. Isbell, T. *Note to Madagascar's Election Winner: Crime, Infrastructure, and Food Insecurity Most Important Issues for Government to Fix*. Afrobarometer Dispatch No. 255. 14 November 2018. Available online: https://afrobarometer.org/sites/default/files/publications/Dispatches/ab_r7_dipatchno255_citizen_priorities_for_madagascar_government.pdf (accessed on 14 October 2019).
119. Huff, A. Weathering the'long wounded year': Livelihoods, nutrition and changing political ecologies in the Mikea Forest Region, Madagascar. *J. Political Ecol.* **2014**, *21*, 83–107. [CrossRef]
120. Fafchamps, M.; Minten, B. Crime, Transitory Poverty, and Isolation: Evidence from Madagascar. *Econ. Dev. Cult. Chang.* **2006**, *54*, 579–603. [CrossRef]
121. Raharimalala, O.; Buttler, A.; Schlaepfer, R.; Gobat, J. Quantifying biomass of secondary forest after slash-and-burn cultivation in Central Menabe, Madagascar. *J. Trop. For. Sci.* **2012**, 474–489.
122. Casse, T.; Milhøj, A.; Ranaivoson, S.; Romuald Randriamanarivo, J. Causes of deforestation in southwestern Madagascar: What do we know? *For. Policy Econ.* **2004**, *6*, 33–48. [CrossRef]
123. Kull, C.A. Madagascar aflame: Landscape burning as peasant protest, resistance, or a resource management tool? *Political Geogr.* **2002**, *21*, 927–953. [CrossRef]
124. Tucker, B.; Huff, A.; Tombo, J.; Hajasoa, P.; Nagnisaha, C. When the wealthy are poor: Poverty explanations and local perspectives in southwestern Madagascar. *Am. Anthropol.* **2011**, *113*, 291–305. [CrossRef]
125. Gardner, C.J.; Davies, Z.G. Rural Bushmeat Consumption Within Multiple-use Protected Areas: Qualitative Evidence from Southwest Madagascar. *Hum. Ecol.* **2014**, *42*, 21–34. [CrossRef]
126. Dirac Ramohavelo, C.; Sorg, J.-P.; Buttler, A.; Reinhard, M. Recommandations pour une agriculture plus écologique respectant les besoins socio-économiques locaux, région du Menabe Central côte ouest de Madagascar. *Madag. Conserv. Dev.* **2014**, *9*, 7. [CrossRef]
127. Filou, E. Illegal corn farming menaces a Madagascar protected area. *Mongabay*, 21 February 2019.
128. Neudert, R.; Goetter, J.F.; Andriamparany, J.N.; Rakotoarisoa, M. Income diversification, wealth, education and well-being in rural south-western Madagascar: Results from the Mahafaly region. *Dev. South. Afr.* **2015**, *32*, 758–784. [CrossRef]
129. Cetas, E.R.; Yasué, M. A systematic review of motivational values and conservation success in and around protected areas. *Conserv. Biol.* **2017**, *31*, 203–212. [CrossRef] [PubMed]

130. Reibelt, L.M.; Woolaver, L.; Moser, G.; Randriamalala, I.H.; Raveloarimalala, L.M.; Ralainasolo, F.B.; Ratsimbazafy, J.; Waeber, P.O. Contact Matters: Local People's Perceptions of Hapalemur alaotrensis and Implications for Conservation. *Int. J. Primatol.* **2017**, *38*, 588–608. [CrossRef]
131. Garnett, S.T.; Burgess, N.D.; Fa, J.E.; Fernández-Llamazares, Á.; Molnár, Z.; Robinson, C.J.; Watson, J.E.M.; Zander, K.K.; Austin, B.; Brondizio, E.S.; et al. A spatial overview of the global importance of Indigenous lands for conservation. *Nat. Sustain.* **2018**, *1*, 369–374. [CrossRef]
132. Fritz-Vietta, N.V.M.; Tahirindraza, H.S.; Stoll-Kleemann, S. Local people's knowledge with regard to land use activities in southwest Madagascar—Conceptual insights for sustainable land management. *J. Environ. Manag.* **2017**, *199*, 126–138. [CrossRef] [PubMed]
133. Thielsen, K. Which form of agreement works for communitybased management? A case study from southwestern Madagascar. *Madag. Conserv. Dev.* **2016**, *11*, 66–77. [CrossRef]
134. Pollini, J.; Hockley, N.; Muttenzer, F.D.; Ramamonjisoa, B.S. The transfer of natural resource management rights to local communities. In *Conservation and Environmental Management in Madagascar*; Routledge: London, UK, 2014; pp. 172–192.
135. Kull, C.A. Empowering Pyromaniacs in Madagascar: Ideology and Legitimacy in Community-Based Natural Resource Management. *Dev. Chang.* **2002**, *33*, 57–78. [CrossRef]
136. Nature. Interdisciplinarity: Scientists must work together to save the world. In *Special Issue: Interdisciplinarity*; Springer Nature Publishing AG.: Berlin, Germany, 2015; Volume 525, p. 7569.
137. Bonebrake, T.C.; Brown, C.J.; Bell, J.D.; Blanchard, J.L.; Chauvenet, A.; Champion, C.; Chen, I.-C.; Clark, T.D.; Colwell, R.K.; Danielsen, F.; et al. Managing consequences of climate-driven species redistribution requires integration of ecology, conservation and social science. *Biol. Rev.* **2018**, *93*, 284–305. [CrossRef]
138. Knight, A.T.; Bode, M.; Fuller, R.A.; Grantham, H.S.; Possingham, H.P.; Watson, J.E.M.; Wilson, K.A. Barometer of Life: More Action, Not More Data. *Science* **2010**, *329*, 141. [CrossRef]
139. Stuart, S.; Wilson, E.; McNeely, J.; Mittermeier, R.; Rodríguez, J. Response to comments on 'Ecology. The barometer of life'. *Science* **2010**, *329*, 141–142. [CrossRef]
140. Cleaver, F. Paradoxes of participation: Questioning participatory approaches to development. *J. Int. Dev.* **1999**, *11*, 597–612. [CrossRef]
141. Teye, J. Corruption and illegal logging in Ghana. *Int. Dev. Plan. Rev.* **2013**, *35*, 1–19. [CrossRef]
142. Starrfmonline. New Madagascar President Rajoelina Impressed with Ghana's Development. Available online: https://starrfm.com.gh/2019/01/new-madagascar-president-rajoelina-impressed-with-ghanas-devt/ (accessed on 13 April 2019).
143. Tollefson, J. Madagascar's forests face uncertain future. *Nature* **2019**, *565*, 407. [CrossRef] [PubMed]
144. Carver, E. Madagascar's next president to take office, bears suspect eco record. *Mongabay*, 18 January 2019.
145. Ratsimbazafy, J.; Jones, J.; Mittermeier, R. Protect Madagascar's national parks from pillage. *Nature* **2019**, *565*, 567. [CrossRef]
146. Jones, J.P.G.; Ratsimbazafy, J.; Ratsifandrihamanana, A.N.; Watson, J.E.M.; Andrianandrasana, H.T.; Cabeza, M.; Cinner, J.E.; Goodman, S.M.; Hawkins, F.; Mittermeier, R.A.; et al. Madagascar: Crime threatens biodiversity. *Science* **2019**, *363*, 825. [CrossRef]
147. Wilme, L.; Waeber, P.O. Madagascar: Guard last of the forests. *Nature* **2019**, *565*, 567.

Article

Relationship between Vegetation and Environment in an Arid-Hot Valley in Southwestern China

Jun Pei, Wei Yang *, Yangpeng Cai, Yujun Yi and Xiaoxiao Li

State Key Laboratory of Water Environment Simulation, School of Environment, Beijing Normal University, Beijing 100875, China; 201621180020@mail.bnu.edu.cn (J.P.); yanpeng.cai@bnu.edu.cn (Y.C.); yiyujun@bnu.edu.cn (Y.Y.); xxli@mail.bnu.edu.cn (X.L.)

* Correspondence: yangwei@bnu.edu.cn; Tel.: +86-159-1118-1787

Received: 25 November 2018; Accepted: 11 December 2018; Published: 14 December 2018

Abstract: The sparse and fragile vegetation in the arid-hot valley is an important indicator of ecosystem health. Understanding the correlation between this vegetation and its environment is vital to the plant restoration. We investigated the differences of soil moisture and fertility in typical vegetation (*Dodonaea viscosa* and *Pinus yunnanensis*) under a range of elevations, slopes, and aspects in an arid-hot valley of China's Jinsha River through field monitoring and multivariate statistical analysis. The soil moisture differed significantly between the dry and rainy seasons, and it was higher at high elevation (>1640 m) and on shade slopes at the end of the dry season. Soil fertility showed little or no variation among the elevations, but was highest at 1380 m. *Dodonaea viscosa* biomass increased, then decreased, with increasing elevation on the shade slopes, but decreased with increasing elevation on the sunny slopes. On the shade slopes, *Pinus yunnanensis* biomass was higher at low elevations (1640 m) than it was on sunny slopes, but lower at high elevation (1940 m) on the sunny slopes. We found both elevation and soil moisture were significantly positively correlated with *P. yunnanensis* biomass and negatively correlated with *D. viscosa* biomass. Thus, changes in soil moisture as a function of elevation control vegetation restoration in the arid-hot valley. Both species are adaptable indigenous plants with good social and ecological benefits, so these results will allow managers to restore the vegetation more effectively.

Keywords: arid-hot valley; vegetation; environmental factors; reservoir region

1. Introduction

The high temperature and low humidity in the arid-hot valley results from a combination of local climatic conditions with strong sunlight and valley wind (particularly winds caused by convection during a forest fire) and evapotranspiration. Poor soil water retention capacity intensifies the difficulty of vegetation growth, leading to serious soil erosion, threatening the human and environment security of Yangtze River basin, especially in the main stream of Jinsha River (the upper Yangtze River) [1]. Since the 1990s, plant introduction and afforestation projects have been carried out in the arid-hot valley in southwest China, but we still do not understand the factors responsible for success or failure of these projects [2].

Exploring the relationship between vegetation and environment is the key to successful vegetation restoration, particularly on mountainous terrain. Many studies have focused on potential impact factors, including climatic factors, topographic conditions, soil properties, and human disturbance [3–5]. A deficiency of soil moisture results in sparse vegetation and decreased restoration effectiveness in the arid-hot valley and, especially at the end of the drought season, the values of soil moisture may be lower than the plant-wilting coefficient [6,7]. In addition, topography plays an important role in vegetation growth and distribution [8], since it affects many aspects of the plant's environment; higher solar radiation and less available water on the sunny slopes of semi-arid valleys would exacerbate drought stress, resulting in the survival of fewer species, a lower plant density, and lower growth rates of the surviving plants [9]. The literature shows that topographic conditions and soil properties are closely related to plant community composition and distribution [10], vegetation cover [11], and biodiversity [12]. To improve ecological restoration efficiency, it is necessary to deeply understand the relationships between the targeted native vegetation and the local environment (e.g., topography, soil, other habitat conditions), and to identify the key factors that affect the growth, survival, and distribution of vegetation in the arid-hot valley [13,14].

The Jinsha River basin in southwestern China has a typical arid-hot valley, and provides the important source of hydroelectric power, and plays a vital role in ecological security for its special ecosystems and rich biodiversity. Since economic development has been driven by the hydroelectric project, dam and reservoir construction have destroyed much local vegetation, thereby exacerbating the contradictions between civilization and nature. Managers of the project recognize this problem, and have looked for ways to promote greener and more sustainable development, including vegetation restoration projects and soil erosion control [14–16]. Many studies on the change of vegetation cover [17], mechanisms and degree of vegetation resistance, soil characteristics and fertility, and restoration techniques and benefits have been carried out [18]. The dominant native plants, *Dodonaea viscosa* and *Pinus yunnanensis*, with high stress resistance and restoration benefit, are important for soil and water conservation, and both are, therefore, key plants in ecological restoration projects. These native plants have attracted much attention from many researchers, but most of the research selected a single factor, such as soil quality, slope and elevation, to statistically analyze its effect on vegetation; even a few studies have focused on more factors, but just discussed the qualitative influence and were carried out only in the tributaries of the Jinsha River [19]. *D. viscosa* biomass is significantly affected by both slopes and aspects [20], and transpiration of *D. viscosa* is strong during both the rainy and dry seasons [21]. *P. yunnanensis* is widely distributed, but grows poorly on sunny slopes, even though it can tolerate low soil moisture and nutrient levels and, especially, a very low phosphorus content [22,23]. However, there has been no integrated analysis of how environmental factors affect these native plants of the Jinsha River valley.

Using the Longkaikou Reservoir region in the Jinsha River basin as a case study, we obtained field data about changes in the characteristics of *D. viscosa* and *P. yunnanensis* across an elevation gradient, including the vegetation biomass and the response to climatic, topographic, and soil characteristics, by means of multiple field surveys to analyze the differences in soil and vegetation as a function of topography. Our goal was to reveal the relationships among the topographic and soil variables and the biomass of *D. viscosa* and *P. yunnanensis*. Our results will provide scientific support for vegetation restoration projects in the arid-hot valley of the Jinsha River.

2. Materials and Methods

2.1. Study Area

Longkaikou hydroelectric station, in Yongsheng County of Yunnan Province, is the sixth of eight cascade reservoirs constructed in the Jinsha River basin. Construction began in 2007, and it began operation in 2014. The reservoir's main purpose is power generation, but it also provides water for irrigation and an urban and industrial water supply. The dam height is 119 m, the normal storage water level is 1298 m asl (above sea level), the backwater length is 41 km, the water area is about 17 ha, and the total reservoir capacity is 5×10^8 m^3. The reservoir is located in a typical alpine canyon area with complex topography and diverse local microclimates. The difference between the rainy and dry seasons is obvious, with an annual average precipitation of 936 mm, of which 82% falls between June and September (Figure S1). Figure S1 shows the annual changes of temperature and precipitation in our study area, based on data from the Yongsheng meteorological station, which is located 37.8 km from our study sites at an elevation of 2151 m asl. The land below an elevation of 1500 m asl has an arid-hot valley climate; above this elevation, there is greater precipitation and the temperatures are much colder. The vegetation, including *D. viscosa*, *P. yunnanensis*, and *Ageratina adenophora*, is widely distributed in the reservoir region, but has been seriously disturbed by construction of the reservoir, as well as by grazing of livestock and cutting. In local vegetation restoration projects, *P. yunnanensis* is used on barren mountain slopes [24]. *D. viscosa* is a perennial evergreen shrub or small tree of the genus *Dodonaea*. It can adapt to a wide range of sites due to its high tolerance of drought and low soil fertility, and is therefore one of the main native plants used for vegetation restoration in the study area. Our study area covers nearly 300 km^2, with a length of 41 km along the river and a width of 7.0 km (Figure 1).

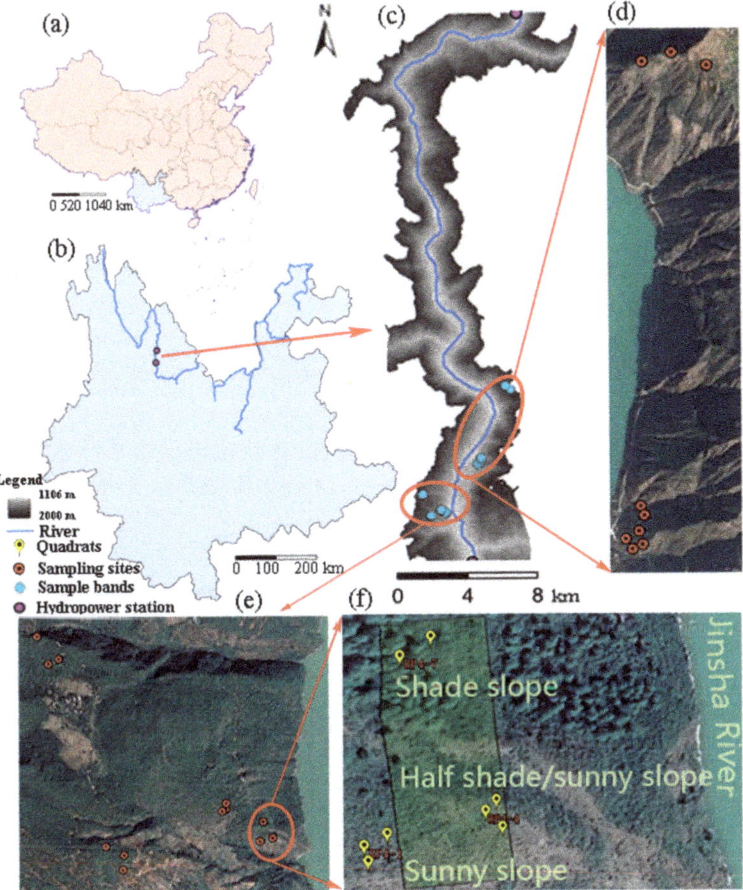

Figure 1. Map of the study region: (**a**) location of Yunnan Province; (**b**) location of the Jinsha River in Yunnan Province; (**c**) map of the reservoir of the Longkaikou Hydropower Station and sample bands; (**d**) and (**e**) locations of the sampling sites; and (**f**) an example of the location of the quadrats at a sampling site.

2.2. Sample Collection and Processing

We conducted field surveys three times (in April and August 2017, and in April 2018) in the study area, and collected soil and vegetation samples during each sampling period. We combined the April data to produce a single dataset for the dry season; the August data represented the end of the rainy season (the rain was too frequent to reach high area with elevation above 1640 m). For each sample, we recorded the elevation, slope, and aspect simultaneously (Table S1). We attempted to establish 3 quadrats at each of 21 sampling sites, for a total of 63 quadrats (*n*); however, some sampling sites had two or four quadrats, due to the nature of the topographic conditions. All of the samples were evenly distributed bath left bank (marked LPx-x) and right bank (marked RPx-x) at five elevations that ranged from the river (ca. 1380 m asl) to the top of the mountains (ca. 1940 m asl), and included shade slopes, sunny slopes, and half-shaded slopes. We established sampling bands at elevations of 1380, 1440, 1520, 1640, and 1940 m asl. For each sampling site, we recorded the slope and aspect (*Asp*) using a DQY-1 compass (Geological Compass; Haerbin, Heilongjiang Province, China), and obtained the elevation and the longitude and latitude using a GPS receive (GPSmap 62sc; Garmin, Lenexa, KS, USA). We calculated the topographic wetness index and the distance from the quadrats to the river are calculated from a digital elevation model (http://www.gscloud.cn/sources/?cdataid=302&pdataid=10) using version 10.2 of the ArcGIS software (www.esri.com).

In total, we established 57 *D. viscosa* quadrats and 38 *P. yunnanensis* quadrats, and recorded number of branches, plant density, and vegetation cover, diameter at breast height, crown width, and height. We collected the aboveground biomass for 25 *D. viscosa* plants, and dried the samples at 60 °C until constant weight in the laboratory; we then weighed the oven-dry biomass using a laboratory electronic balance with a precision of 0.01 g.

We collected 64 soil samples (each ca. 1 kg) to a depth of 10 cm with a shovel, as the soil was too rocky below this depth to allow sampling. We measured soil moisture, organic matter, total nitrogen, total potassium, hydrolyzable nitrogen, and available phosphorus in the laboratory according to the standards of the Chinese Forestry Bureau (http://www.zbgb.org/StandardCList25C.htm) using the air-dry soil for all parameters except soil moisture. Soil moisture was measured based on the difference between the fresh and oven-dry mass (after drying at 105 °C for 24 h until all the moisture was driven off). The organic matter was determined by the potassium dichromate oxidation method with external heating (LY/T 1237-1999), total nitrogen was measured by the Kjeldahl method (LY/T 1228-2015), total potassium was determined by NaoH flame photometry (LY/T 1234-2015), hydrolyzable nitrogen was determined by the alkaline hydrolysis-diffusion method (LY/T 1228-2015), and available phosphorus was determined by colorimetry (LY/T 1232-2015).

2.3. Estimation of Soil Fertility and Vegetation Biomass

We established a holistic index of soil fertility (*SF*) as follows:

$$SF = \sum_{i=1}^{5} \frac{x_i - x_{\mathrm{min}i}}{x_{\mathrm{min}i}}, \tag{1}$$

where x_i represents the value of nutrient indicator *i* (organic matter, total nitrogen, total potassium, hydrolyzable nitrogen, and available phosphorus), and $x_{\mathrm{min}i}$ is the lowest value for each of the five nutrient indicators in the soil nutrient grading standards of the second national land survey.

The growth model for *D. viscosa* is region-dependent, since this species lives in different areas [25], so we established a relative growth model.

Due to the multiple branches produced by this shrub, we defined the total diameter (*D*) based on the sum of the squared branch diameters:

$$D = \sqrt{\sum_{i=1}^{n} D_i^2}, \tag{2}$$

where D_i is the diameter (cm) of branch *i*, and *n* is the number of branches contained in each shrub [26]. We then adopted a relative growth model in the form of a power function based on a previously determined growth model for the shrub [27]:

$$Bio = 44.047\ D^2\ H^{0.467}\ (R^2 = 0.94, p < 0.01), \quad (3)$$

where *Bio* = aboveground biomass, *D* is the total diameter from Equation (2), and *H* is the height at the end of the tallest branch.

We obtained an equation from Huo et al. [28] to determine the relative growth of *P. yunnanensis*.

$$Bio = 0.026\ D^{2.83} \quad (4)$$

2.4. Statistical Analysis

We used cosine function convert aspect (ASP) and divided it into shade slope (<0) and sunny slope (>0). First of all, we have a Kolmogorov–Smirnov test for normality of the data. When the data satisfied the conditions of normal distribution, we used ANOVA and LSD to identify significant differences in vegetation and soil properties among the topographic conditions. We used regression analysis to establish the relative growth models to calculate plant biomass and used Pearson's correlation coefficient (*r*) to quantify the strength of the relationships among the vegetation, topographic, and soil indicators. All the statistical analysis was performed using version 3.5.1 of the R software (www.r-project.org).

We used redundancy analysis (RDA) to study the relationships between the vegetation and environmental indexes because the eigenvalue of the first axis of a detrended correspondence analysis (DCA) was less than 3 for the vegetation data. We performed these analyses using the "vegan" package for the R software (https://cran.r-project.org/web/packages/vegan/index.html).

3. Results

3.1. Variation in Soil Moisture and Fertility

Figure 2 shows the changes of soil moisture along the elevation gradient. The soil moisture was higher on the shade slopes than on the sunny slopes at all elevations in the dry season (Figure 2a), and it was much higher in the rainy season than at the end of the dry season at a given elevation (Figure 2b). The soil moisture increased slightly, but not significantly, with increasing elevation, except the shade slope at 1940 m asl ($p < 0.05$) and, at elevations ($\leq$1520 m), with an average of 0.055, did not differ significantly among elevations.

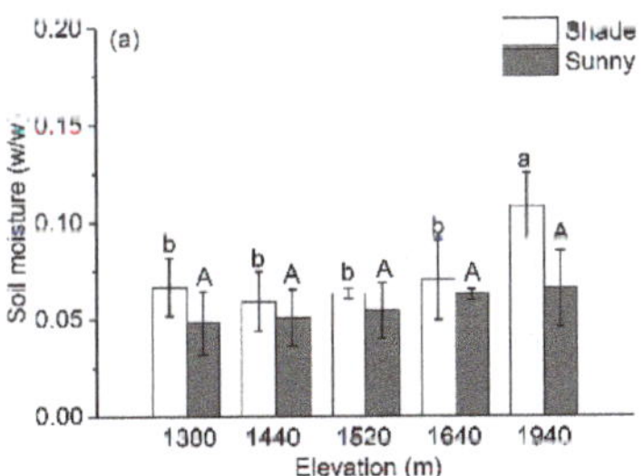

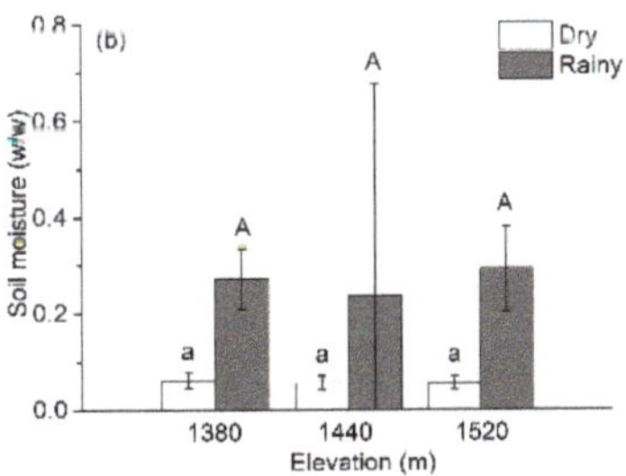

Figure 2. Changes in soil moisture (**a**) along the elevation gradient for shade and sunny sites and (**b**) between seasons at each elevation. Values are mean ± SD. Note that the values on the y-axis differ greatly between the two graphs. Values labeled with different lowercase letters differ significantly (*LSD*, $p < 0.05$) between elevations for a given site category (sunny and rainy); values labeled with the same capital letters did not differ significantly between elevations for a given site category (shade and dry).

There was no significant difference in soil fertility (based on the holistic index of soil fertility) among the elevations (Figure 3). The soil fertility reached its maximum value (49.07) at the lowest elevation (1380 m) on the sunny slope, and its minimum value (19.57) at 1520 m on the shade slope.

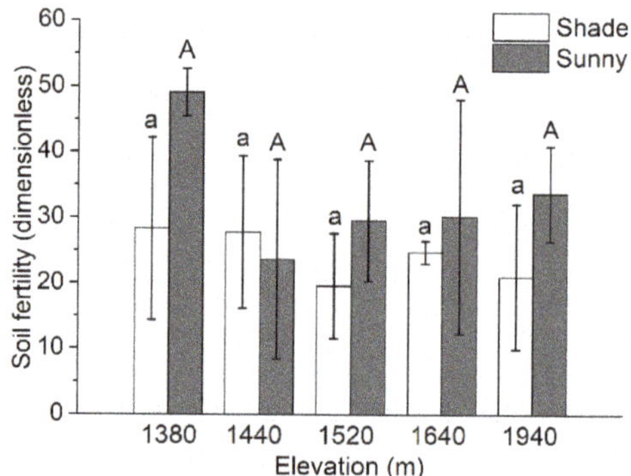

Figure 3. Changes in soil fertility (based on the holistic index of soil fertility; Equation (1)) with increasing elevation for the aspects. Values are mean ± SD. Values labeled with same lowercase letters differ not significantly (*LSD*, $p > 0.05$) between elevations for the shade slope; values labeled with the same capital letters did not differ significantly between elevations on the sunny slope.

3.2. Changes in Vegetation Biomass as a Function of Elevation and Aspect

Table 1 summarizes the changes in the characteristics of the vegetation as a function of elevation. For *D. viscosa*, the density and crown width increased with increasing elevation, reaching a significantly higher maximum at 1520 m, then decreased thereafter. Diameter at breast height did not differ significantly among elevations, except for a significant decrease at 1940 m. Height was significantly higher at the two lowest elevations than at the highest elevations. All *D. viscosa* parameters presented their lowest values at 1940 m. For *P. yunnanensis*, density increased with increasing elevation, and the differences were generally significant. Diameter at breast height, crown width, and height showed a similar pattern of increase, but with a maximum at 1640 m. Vegetation cover showed an inconsistent pattern, but reached a significantly higher maximum value at 1520 m. The density of the *P. yunnanensis* increased from 0.1 to 13.0 per 100 m^2, and the average diameter, crown width, and height at 1940 increased to approximately 4 to 10 times the values at 1380 m.

Table 1. Vegetation survey data for the two species along an elevation gradient in the study area. Values are mean ± SD. For a given parameter and species, values labeled with different letters differed significantly between elevations (*LSD*, $p < 0.05$).

Species	Elevation (m asl)	Density (n./m²)	Diameter at Breast Height [a] (cm)	Crown Width (cm)	Height (cm)	Vegetation Cover (%)
Dodonaea viscosa	1380	5.0 ± 2 b	2.44 ± 1.16 a	96 ± 32.97 a	179 ± 24.55 a	46 ± 27.25 ac
	1440	5.0 ± 3.43 b	2.47 ± 1.34 a	100 ± 41.82 a	183 ± 53.20 a	33 ± 10.17 b
	1520	12.0 ± 6.27 a	2.31 ± 0.50 a	103 ± 20.88 a	151 ± 29.74 ab	58 ± 20.17 a
	1640	7.0 ± 2.10 b	1.61 ± 0.56 ab	95 ± 34.82 a	142 ± 16.93 ab	40 ± 10.95 ab
	1940	4.0 ± 4.57 b	0.75 ± 0.54 b	60 ± 62.57 a	95 ± 56.72 b	31 ± 18.03 d
Pinus Yunnanensis	1380	0.1 ± 0.33 D	2.62 ± 7.86 A	39 ± 116.67 D	78 ± 233.33 D	46 ± 27.25 AB
	1440	1.0 ± 1.39 D	10.21 ± 13.66 A	180 ± 233.67 B	309 ± 372.03 CD	33 ± 10.17 C
	1520	3.0 ± 2.97 BD	11.19 ± 9.50 A	278 ± 248.72 AC	481 ± 403.50 BC	66 ± 21.78 A
	1640	11.0 ± 5.39 AC	11.47 ± 3.35 A	400 ± 57.01 A	875 ± 183.71 A	40 ± 10.95 B
	1940	13.0 ± 8.56 A	10.88 ± 3.65 A	287 ± 102.44 A	821 ± 350.34 AB	31 ± 18.03 C

[a] Cumulative diameter of all stems for *D. viscosa*. Values labeled with same lowercase letters differ not significantly (*LSD*, $p > 0.05$) between elevations for *D. viscosa*; values labeled with the same capital letters did not differ significantly between elevations for *P. yunnanensis*.

Figure 4 presents the aboveground biomass of the two species. The biomass of *D. viscosa* first increased, and then decreased with increasing elevation on the shade slopes, with a maximum of 0.74 kg/m^2 at 1440 m; by contrast, it decreased with increasing elevation on the sunny slopes, with a maximum value of 0.75 kg/m^2 at 1380 m (Figure 4a). The biomass of *P. yunnanensis* biomass first increased and then decreased with increasing elevation on the shade slope, with a maximum value of 5.87 kg/m^2 at 1520 m, but increased with increasing elevation on the sunny slopes after it first appeared at an elevation of 1520 m (Figure 4b).

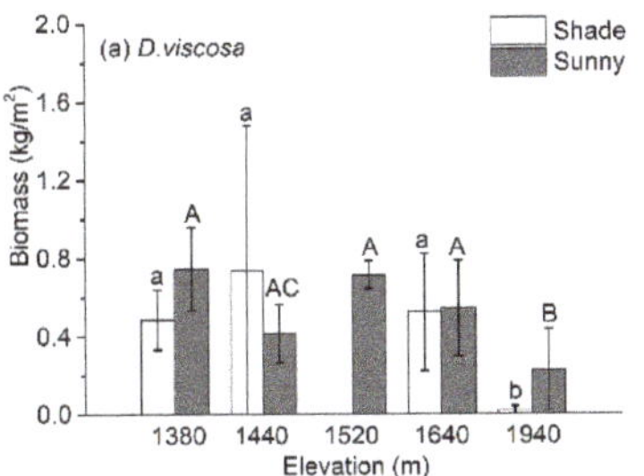

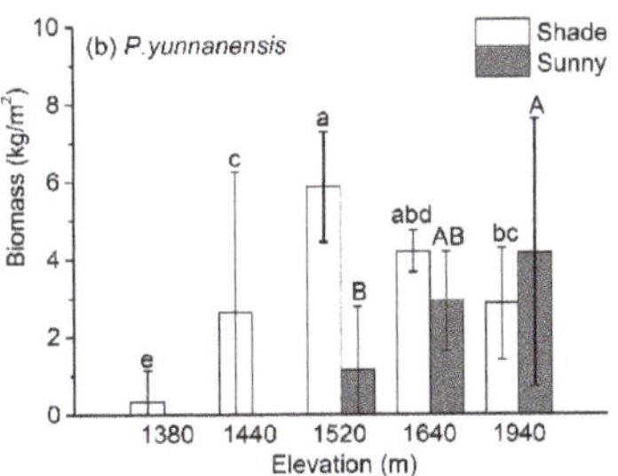

Figure 4. Changes in the aboveground biomass of (**a**) *D. viscosa* and (**b**) *P. yunnanensis* as a function of elevation. Values are mean ± SD. Note that the values on the y-axis differ greatly between the two graphs. Values labeled with different lowercase letters differ significantly (*LSD*, $p < 0.05$) between elevations for a given site category (shade slope); values labeled with same capital letters did not differ significantly between elevations for the sunny slope).

3.3. Relationship between Vegetation and Environment

We used Pearson's correlation coefficient (r) to reveal the significant ($p < 0.05$) relationships between the vegetation and environmental factors (Supplemental Table S2). For *D. viscosa*, biomass was positively correlated with branches, density, diameter at breast height, height, and hydrolyzable nitrogen, but negatively correlated with elevation and distance of the quadrat from the river. Branches were positively correlated with density. Diameter at breast height was positively correlated with crown width and height, but negatively correlated with elevation, distance of quadrat from the river, and soil moisture. Crown width was positively correlated with height. Height was negatively correlated with elevation, distance of quadrat from the river, and soil moisture. For *P. yunnanensis*, biomass was positively correlated with density, diameter at breast height, crown width, height, elevation, and distance of the quadrat from the river. Density was positively correlated with height, elevation, distance of the quadrat from the river, soil moisture, and organic matter. Diameter at breast height was positively correlated with crown width, height, and aspects. Crown width was positively correlated with height. Height was positively correlated with elevation, distance of the quadrat from the river, and soil moisture. Vegetation cover was positively correlated with the topographic wetness index.

Among the significant topographic factors, elevation was positively correlated with the distance of the quadrat from the river and soil moisture; and distance of the quadrat from the river was positively correlated with soil moisture and organic matter. Among the significant soil factors, total nitrogen was positively correlated with available nitrogen and soil fertility, and the available nitrogen and phosphorus were both positively correlated with soil fertility.

In addition, we used redundancy analysis to clarify the relationships among the soil properties and topographic parameters for the two species. Figures 5 and 6 present the results for *D. viscosa* and *P. yunnanensis*, respectively. When the vegetation parameters were used as response variables, and the soil properties and topographic parameters were used as independent variables, RDA analysis (Figure 5) showed that the constrained variables explained 64% of the total variance for *D. viscosa* along all RDA axes. The distance from quadrats to the river, soil moisture, and elevation were positively correlated and contributed strongly to RDA axis 1; soil moisture was mainly affected by elevation and distance from quadrats to the river, which agrees with the correlation results (Table S2). The biomass, height, crown width, and diameter at breast height were negatively correlated with soil moisture, whereas the slope and total nitrogen were positively correlated with *D. viscosa* biomass, and the topographic wetness index and total potassium significantly affected the density and vegetation cover of *D. viscosa*. We divided the environmental variables into two independent variables: soil properties and topographic conditions. The soil properties explained 30% of the total variation of *D. viscosa*, and topographic conditions explained 51%.

The RDA analysis for *P. yunnanensis* (Figure 6) showed that the constrained variables explained 69% of the total variance along all RDA axes. The distance from quadrats to the river, soil moisture, and elevation were positively correlated and contributed strongly to the first RDA axis. The vegetation parameters were positively correlated with soil moisture. In addition, the slope aspect and available phosphorus were positively correlated with the biomass of *P. yunnanensis*, while the total nitrogen was negatively correlated with the vegetation characteristics. The RDA analysis showed that soil properties explained 31% of the variation and topographic variables explained 53%.

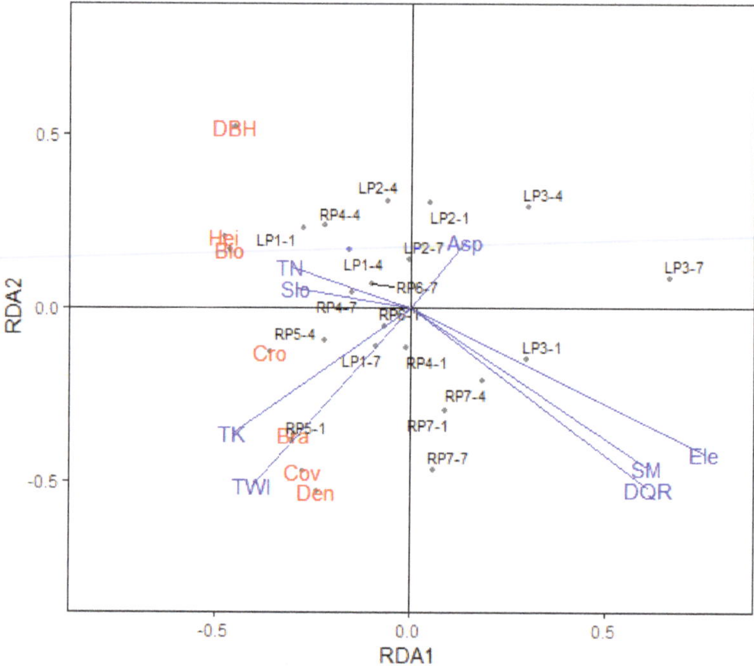

Figure 5. Results of the redundancy analysis for the relationships between the vegetation parameters for *D. viscosa*, soil properties, and topographic conditions. Variable names: AP, available phosphorus; Bio, aboveground biomass; Bra, branches; Cov, vegetation cover; Cro, crown width; DBH, diameter at breast height; Den, plant density; DQR, distance from quadrats to the river; Ele, elevation; Hei, height; HN, hydrolyzable nitrogen; ASP, aspect; OM, soil organic matter; Slo, slope; SM, soil moisture; TK, total potassium; TN, total nitrogen; TWI, topographic wetness index.

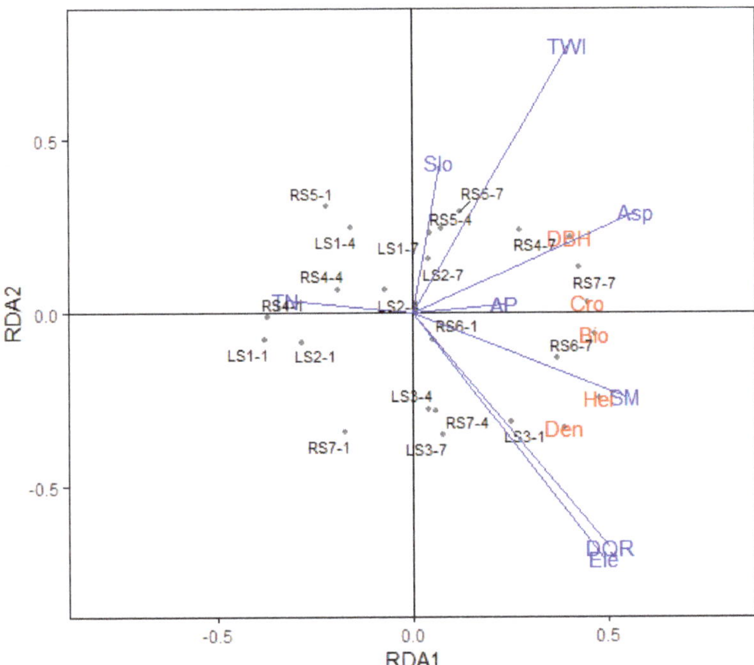

Figure 6. Results of the redundancy analysis for the relationships between the vegetation parameters for *P. yunnanensis* and the soil properties and topographic conditions (the abbreviations in Figure 6 was the same as them in Figure 5).

4. Discussion

4.1. Effects of Habitat Conditions on Vegetation

The arid-hot valley has fragile habitats because of the highly variable spatial and temporal distribution of precipitation, and barren nutrient conditions. Our results showed large differences in soil moisture between the rainy and dry seasons, between the shade and sunny slopes, and between elevations, which is consistent with previous research results [6,29] The soil moisture at an elevation (≤1520 m) averaged 0.055, and was below the wilting point of 0.059 for *D. viscosa* [30], possibly because the continuously intense evaporation during the dry season leads to low soil moisture [31]. Soil moisture was higher at 1380 m than 1440 m, and soil fertility was highest at the lowest elevation (1380 m), probably due to the increased relative humidity near the river. In fact, increased relative humidity and livestock grazing at lower elevations would accelerate the decomposition and accumulation of nutrients in the soil [32]. Our results also show differences in vegetation biomass and morphology along the elevation gradient, though the two species showed different patterns: *D. viscosa* generally either increased in size and then decreased with increasing elevation, or decreased steadily, whereas *P. yunnanensis* tended to increase in size with increasing elevation.

It will important to build on the present results to improve our understanding of the key factors that affect the growth, development, and distribution of organisms along the elevation gradient in our study area. The complex and diverse relationships between soil properties and topographic conditions affect the composition and biodiversity of the regional vegetation community [3,33], and provide insights into the characteristics that will make a plant suitable for local vegetation restoration.

Our RDA results suggest that topographic conditions explain more of the vegetation variation than soil properties, especially for elevation. This results from the high mountains and deep valleys in the Jinsha River basin. Rossi et al. [34] showed that topographic conditions, and especially the elevation and slope, could explain local vegetation patterns well in high mountain areas. The significant correlations among the elevation, distance from quadrats to the river, and soil moisture, and their significant effects on the biomass and height of *D. viscosa* and *P. yunnanensis*, confirmed that elevation was the dominant factor that affected vegetation growth and distribution in our study area. Elevation is often regarded as a comprehensive physical index that integrates the effects of changes in precipitation, temperature, relative humidity, and other environmental conditions, all of which affect soil properties and vegetation growth. Close relationships among the elevation, soil moisture, and vegetation were also found in China's Loess Plateau [11]. The topographic wetness index has also been strongly correlated with vegetation cover, indicating that it can reflect the overall growth of the vegetation community through its ability to account for microtopographic water distribution and perhaps salinity redistribution [35]. *D. viscosa* and *P. yunnanensis* are widely distributed in our study area, due to their strong tolerance for drought barren soil, so the correlation was weak between vegetation parameters and soil nutrients in our study region.

The decreased precipitation and intense evaporation in the dry season exacerbate the effects of drought stress at low elevations, and led to little or no growth of *P. yunnanensis*, but sparse growth of *D. viscosa* on the sunny slopes. The slightly higher soil moisture on the shade slopes could be crucial to help plants survive the drought period, and this may be the main reason for differences in the plant community and ecological landscape in the arid-hot valley [29]. At an elevation of 1640 m, the growth of *D. viscosa* decreased, whereas *P. yunnanensis* responded strongly to the obviously increased soil moisture at that elevation, suggesting that the increase of soil moisture, combined with the lower temperature at that elevation, was more beneficial for the growth of *P. yunnanensis*. Xiong et al. [36] thought that the geotechnical properties of the site, such as infiltration and lithology, determined the soil moisture and vegetation types in arid-hot valley, but thought that elevation was not important. However, the results of our RDA analyses (Figures 5 and 6) showed that elevation and distance from the quadrats to the river provide a comprehensive index that reflects the change of community characteristics, especially for biomass. Thus, understanding the soil moisture conditions at each site will be a key to successful vegetation restoration.

4.2. Restoration Potential of *P. yunnanensis* and *D. viscosa*

Different plant types have different adaptation mechanisms that determine how they respond to environmental stress. Under drought stress, *D. viscosa* decreases its net photosynthetic rate and improves its water-use efficiency [37], whereas *P. yunnanensis* distributes more photosynthate to root organs and reduces water consumption by stems and leaves [38]. Thus, *D. viscosa* can still survive when soil moisture is extremely low. However, when the two species coexist under suitable soil moisture levels, *P. yunnanensis* grows more vigorously, and its shade may limit the growth of *D. viscosa*, in which the overhead shading lead to the decreasing of *D. viscosa* biomass and increasing of soil moisture. Based on these different responses, it should be possible to select the most appropriate species according to soil moisture conditions at each site where vegetation restoration will occur. Without performing this analysis, the restoration may fail or achieve poor results, and if the species uses more water than the site can sustainably provide, this may exacerbate soil drought [39,40]. Elevation was negatively correlated with *D. viscosa* growth, but positively correlated with *P. yunnanensis* growth, indicating that drought-tolerant *D. viscosa* should be restored at the lower elevations and on sunny slopes, where drought stress is most likely to be severe. The positive correlation between plant parameters for slopes suggests that a suitable slope is beneficial for the growth of *D. viscosa*. *P. yunnanensis* was not sensitive to slope, but its growth was negatively correlated with total nitrogen and positively correlated with available phosphorus. Controlled experiments found that *P. yunnanensis* seedlings could adapt well to nitrogen stress, but were obviously affected by a phosphorus deficiency [41], but a field investigation showed that nitrogen and phosphorus were positively correlated with the growth characteristics of *P. yunnanensis*, with the greatest constraint created by low nitrogen [42].

Although response mechanisms may be different, we concluded that soil nutrients is important for *P. yunnanensis* restoration.

For successful vegetation restoration in the arid-hot valley, we should focus on which factors will most strongly constrain survival and growth; our results and previous research suggest that soil moisture will be critical. *D. viscosa* is highly resistant to drought stress, and also develops considerable biomass, so it is a promising species for vegetation restoration in arid and rocky sites, such as those in our study area. In addition, it has ornamental and medicinal value [43,44]. Some experiments in arid-hot valley have shown that *D. viscosa* grows fast, sprouts many tillers, and can restore a degraded site rapidly [45]. However, *D. viscosa* may be unable to survive drought stress combined with competition from other species and damage caused by grazing, especially at the seedling stage; thus, it requires suitable protection, such as clear weeds, moderate grazing, and increased soil moisture using a water-retaining agent [46], to ensure its survival and growth. *P. yunnanensis* is also widely distributed in our study area, and has produced an obvious improvement of water yields, thereby reducing soil erosion and improving water storage, which together can improve soil carbon storage [47]. However, it is less able than *D. viscosa* to withstand drought and poor soil fertility, so the survival and growth of *P. yunnanensis* plantations could be improved by soil management supplying sufficient nitrogen and phosphorus to mitigate any soil limitations, combined with scientific planting at an appropriate density, which agrees with previous recommendations [19]. In actual vegetation restoration, we must also consider the economic suitability and maintenance needs of the selected resistant plants, which could be considered in our future researches.

Through sampling and analysis, we have improved our understanding of the relationships among vegetation, topographic, and soil properties in the arid-hot valley. In the future, additional work should be carried out to build on our findings. First, we performed our field study only during two typical seasons. Continuous monitoring of surface soil moisture throughout the year, along with changes in plant physiology, such as the degree of water stress, would provide a clearer understanding of its variation and its effects on plants. In addition, we based the relationship between vegetation and environmental conditions mainly on correlations in field data, rather than using controlled experiments that would allow a detailed exploration of the underlying mechanisms. Future research should focus on such experiments to clarify the underlying processes that define the relationships between the vegetation and environmental factors. Finally, it will be necessary to understand the water balance in the study ecosystem based on the relationships among plants, topography, and soil, so that we can choose suitable restoration species for each combination of these conditions, and improve the likelihood of successful restoration.

5. Conclusions

In this study, we surveyed the vegetation parameters, topographic variables, and soil properties in the dry and rainy seasons from 2017 to 2018 in the arid-hot valley in southwestern China, and analyzed the relationships among them by using Pearson's correlation coefficient and redundancy analysis. Our results suggest that soil moisture was relatively adequate during the rainy season, but that only *D. viscosa* survived on sunny slopes at lower elevations ($\leq$1520 m) near the end of the dry season because of the low average soil moisture. The lack of large differences in soil fertility among the elevations suggest that soil fertility would have a relatively small influence on vegetation restoration. The biomass of *D. viscosa* on the shade slopes initially increased with increasing elevation, then decreased again as conditions became unsuitable for the species, but decreased steadily with increasing elevation on the sunny slopes. By contrast, the biomass of *P. yunnanensis* at low elevations ($\leq$1640 m) was higher on the shade slopes than on the sunny slopes, but it was less at high elevation (1940 m). RDA analysis showed that topographic conditions explained vegetation variation better than soil conditions for both species, but that changes in soil moisture along the elevation gradient at the end of the dry season strongly affected vegetation restoration. Our study provides important scientific support for planning ecological restoration in our study area.

Supplementary Materials: The following are available online at http://www.mdpi.com/2071-1050/10/12/4774/s1, Figure S1: Distribution of temperatures and precipitation for Yongsheng County meteorological station. Variable names: Jan, January; Feb, February; Mar, March; Apr, April; Jun, June; Jul, July; Aug, August; Sep, September; Oct, October; Nov, November; Dec, December, Table S1: Topography conditions and soil nutrient elements of quadrats in study area, Table S2: Correlation analysis (Pearson's r) for the relationships among the vegetation and environmental factors for *Dodonaea viscosa* and *Pinus yunnanensis*. Significance: *, $P < 0.05$; **, $P < 0.01$. Bio: Biomass; Den: density; Bra: branches; Cro: crown; Hei: height; Cov: coverage; Asp: aspect slope; Slo: slope; Ele: elevation; DBH: diameter at breast height; DQR: distance from quadrates to river; TWI: topographic wetness index; SM: soil moisture; OM: organic matter; TN: total nitrogen; TK: total potassium; AN: hydrolyzed nitrogen; AP: available phosphorus; SF: soil fertility.

Author Contributions: Conceptualization, W.Y.; Data curation, W.Y.; Formal analysis, W.Y.; Funding acquisition, Y.C.; Investigation, J.P., Y.Y. and X.L.; Methodology, J.P. and X.L.; Project administration, Y.C.; Resources, Y.C.; Software, J.P. and X.L.; Supervision, W.Y.; Validation, Y.Y.; Writing—original draft, J.P.; Writing—review & editing, W.Y. and Y.Y.

Funding: National Key Research and Development Plan of China (No. 2016YFC0502209 and 2017YFC0404505), and the National Natural Science Foundation of China (No. 51579012).

Acknowledgments: We thank the National Key Research and Development Plan of China (No. 2016YFC0502209 and 2017YFC0404505), the Beijing Nova Program (No. Z171100001117080), and the National Natural Science Foundation of China (No. 51579012) for their financial support. We also thank Geoffrey Hart for providing language help during the writing of this paper.

Conflicts of Interest: The authors declare no conflicts of interest.

References

1. Jiang, L.G.; Yao, Z.J.; Liu, Z.F.; Wu, S.S.; Wang, R.; Wang, L. Estimation of soil erosion in some sections of Lower Jinsha River based on RUSLE. *Nat. Hazards* **2015**, *76*, 1831–1847. [CrossRef]
2. Ren, F.P.; Zhang, P.C.; Chen, X.P.; Ping, X.; Chen, J. Vegetation spatial heterogeneity of valleys along the Jinsha River and its influence on ecological restoration. *J. Yangtze River Sci. Res. Inst.* **2016**, *33*, 24–29. (In Chinese)
3. Bennie, J.; Hill, M.O.; Baxter, R.; Huntley, B. Influence of slope and aspect on long-term vegetation change in British chalk grasslands. *J. Ecol.* **2006**, *94*, 355–368. [CrossRef]
4. Solon, J.; Degórski, M.; Roo-Zielińska, E. Vegetation response to a topographical-soil gradient. *Catena* **2007**, *71*, 309–320. [CrossRef]
5. Li, C.; Xiao, B.; Wang, Q.H.; Zheng, R.L.; Wu, J.Y. Responses of soil seed bank and vegetation to the increasing intensity of human disturbance in a semi-arid region of northern China. *Sustainability* **2017**, *9*, 1837. [CrossRef]
6. Mu, J.; Li, Z.B.; Li, P.; Xue, S.; Zhen, Y. Study on soil moisture dynamic variation law in dry seasons in dry-hot valley areas. *J. Yangtze River Sci. Res. Inst.* **2009**, *26*, 22–25. (In Chinese)
7. Lang, N.J. Study on the Influencing Factors of Vegetation Restoration in Degraded Ecological System of Dry-Hot Valleys in Yunnan Province of China. Ph.D. Thesis, Beijing Forestry University, Beijing, China, 2005.
8. Davies, R.G.; Orme, C.D.L.; Storch, D.; Olson, V.A.; Thomas, G.H.; Ross, S.G.; Ding, T.S.; Rasmussen, P.C.; Bennett, P.M.; Owens, I.P. Topography, energy and the global distribution of bird species richness. *Proc. R. Soc. B-Biol. Sci.* **2007**, *274*, 1189–1197. [CrossRef] [PubMed]
9. Måren, I.E.; Karki, S.; Prajapati, C.; Yadav, R.K.; Shrestha, B.B. Facing north or south: Does slope aspect impact forest stand characteristics and soil properties in a semiarid trans-Himalayan valley? *J. Arid Environ.* **2015**, *121*, 112–123. [CrossRef]
10. Poelking, E.L.; Schaefer, C.E.R.; Fernandes Filho, E.I.; De Andrade, A.M.; Spielmann, A.A. Soil-landform-plant-community relationships of a periglacial landscape on Potter Peninsula, maritime Antarctica. *Solid Earth* **2015**, *6*, 583–594. [CrossRef]
11. Hu, X.M.; Cheng, J.M.; Wan, H.E.; Zhao, Y.Y. Reciprocal relationships between topography, soil moisture, and native vegetation patterns in the loess hilly region, China. *Acta Ecol. Sin.* **2006**, *26*, 3276–3285.
12. Stein, A.; Gerstner, K.; Kreft, H. Environmental heterogeneity as a universal driver of species richness across taxa, biomes and spatial scales. *Ecol. Lett.* **2014**, *17*, 866–880. [CrossRef] [PubMed]
13. Song, C.J.; Qu, L.Y.; Tian, Y.S.; Ma, K.M.; Zao, L.X.; Xu, X.L. Impact of multiple soil nutrients on distribution patterns of shrubs in an arid valley, in Southwest China. *Pak. J. Bot.* **2014**, *46*, 1621–1629.

14. Li, K.; Liu, F.Y.; Yang, Z.Y.; Sun, Y.Y. Study status and trends of vegetation restoration of dry-hot valley in Southwest China. *World For. Res.* **2011**, *24*, 55–60. (In Chinese)
15. Wang, C. Research conception of ecological protection and restoration of high dams and large reservoirs construction and hydropower cascade development in southwestern China. *Adv. Eng. Sci.* **2017**, *49*, 19–26. (In Chinese)
16. Kang, H.; Um, M.J.; Park, D. Assessing the habitat suitability of dam reservoirs: A quantitative model and case study of the Hantan River Dam, South Korea. *Sustainability* **2016**, *8*, 1117. [CrossRef]
17. Zhang, B.; Wang, D.; Wang, G.G.; Ma, Q.; Zhang, G.B.; Ji, D.M. Vegetation cover change over the southwest china and its relation to climatic factors. *Res. Environ. Yangtze Basin* **2015**, *24*, 956–964. (In Chinese)
18. Yang, J.D.; Zhang, Z.M.; Shen, Z.H.; Ou, X.K.; Geng, Y.P.; Yang, M.Y. Review of research on the vegetation and environment of dry-hot valleys in Yunnan. *Biodivers. Sci.* **2016**, *24*, 462–474. (In Chinese) [CrossRef]
19. Li, Q. Researches on Eco-Environment Features and Vegetation Restoration Technologies in Dry-Hot Valley of Jinsha River. Master's Thesis, Xi'an University Technology, Xi'an, China, 2008. (In Chinese)
20. Huang, D.; Ge, L.; Ma, H.C.; Zhao, G.J.; Yang, J.J. Growth response of *Dodonaea viscosa* L. to different site factors in the hot regions of Yunnan Province. *J. Northeast For. Univ.* **2015**, *6*, 22–24. (In Chinese)
21. Li, K.; Zeng, J.M. A study on transpiration of some tree species planted in hot and arid valley of Jinsha River. *For. Res.* **1999**, *12*, 244–250. (In Chinese)
22. Deng, X.Q.; Huang, B.L.; Wen, Q.Z.; Hua, C.L.; Tao, J. A research on the distribution of *Pinus yunnanensis* forest in Yunnan Province. *J. Yunnan Univ. (Nat. Sci. Ed.)* **2013**, *35*, 843–848. (In Chinese)
23. Cai, N.H.; Zhang, R.L.; Chen, S.; Bai, Q.S.; Xu, Y.L. The physiological responses of *Pinus yunnanensis* to water stress. *J. Yunnan Agric. Univ.* **2013**, *2*, 247–250. (In Chinese)
24. Dai, K.J.; He, F.; Shen, Y.X.; Zhou, W.J.; Li, Y.P.; Tang, L. Advances in the research on *Pinus yunnanensis* forest. *J. Central South For. Univ.* **2006**, *2*, 138–142. (In Chinese)
25. Litton, C.M.; Boone, K.J. Allometric models for predicting aboveground biomass in two widespread woody plants in Hawaii. *Biotropica* **2008**, *40*, 313–320. [CrossRef]
26. Beets, P.N.; Kimberley, M.O.; Oliver, G.R.; Pearce, S.H. The application of stem analysis methods to estimate carbon sequestration in arboreal shrubs from a single measurement of field plots. *Forests* **2014**, *5*, 919–935. [CrossRef]
27. Zeng, W.S. A review of studies of shrub biomass modeling. *World For. Res.* **2015**, *28*, 31–36. (In Chinese)
28. Huo, C.F.; Lu, X.Y.; Cheng, G.W. Parameter identification and simulation verifying of southwestern forest succession model. *J. Northeast For. Univ.* **2012**, *40*, 78–83. (In Chinese)
29. Jiang, J.M.; Fei, S.M.; Wang, P.; Lei, C.H.; He, X.F. Dynamic comparison of soil water between shady slopes and sunny slopes in the dry-hot valley. *J. Sichuan For. Sci. Technol.* **2005**, *26*, 30–35. (In Chinese)
30. Jiang, J.M.; Fei, S.M.; Li, H.; Lei, C.H. Comparison of drought resistance ability of several main afforestation trees in Panzhihua dry-hot valley. *Trans. China Pulp. Paper* **2004**, 345–348. (In Chinese)
31. Dong, Y.F.; Xiong, D.H.; Su, Z.A.; Li, J.J.; Yang, D.; Shi, L.T.; Liu, G.C. The distribution of and factors influencing the vegetation in a gully in the dry-hot valley of southwest China. *Catena* **2014**, *116*, 60–67. [CrossRef]
32. Zheng, Y.; Li, Z.B.; Li, P.; Mu, J. Features of land use type on soil properties in the dry-hot valley of Jinsha River. *Res. Soil Water Conserv.* **2010**, *17*, 174–177. (In Chinese)
33. Moeslund, J.E.; Arge, L.; Bøcher, P.K.; Dalgaard, T.; Odgaard, M.V.; Nygaard, B.; Svenning, J.C. Topographically controlled soil moisture is the primary driver of local vegetation patterns across a lowland region. *Ecosphere* **2013**, *4*, 1–26. [CrossRef]
34. Rossi, G.; Ferrarini, A.; Dowgiallo, G.; Carton, A.; Gentili, R.; Tomaselli, M. Detecting complex relations among vegetation, soil and geomorphology. An in-depth method applied to a case study in the Apennines (Italy). *Ecol. Complex.* **2014**, *17*, 87–98. [CrossRef]
35. Wu, D.Q.; Liu, J.; Wang, W.; Ding, W.J.; Wang, R.Q. Multiscale analysis of vegetation index and topographic variables in the Yellow River Delta of China. *J. Plant Ecol.* **2009**, *33*, 237–245. (In Chinese)
36. Xiong, D.H.; Zhou, H.Y.; Yang, Z.; Zhang, X.B. Slope lithologic property, soil moisture condition and revegetation in dry-hot valley of Jinsha River. *Chin. Geogr. Sci.* **2005**, *15*, 186–192. (In Chinese) [CrossRef]
37. Li, J.Y.; Jia, L.Q.; Lang, N.J.; Chen, S.Y.; Wu, L.Y. Characteristics of photosynthesis of *Dodonaea viscosa* in the dry-hot valley of Jinsha River. *J. Beijing For. Univ.* **2003**, *25*, 20–24. (In Chinese)

38. Duan, X.; Zhao, Y.Y. Growth and biomass of *Pinus yunnanensis* seedlings under water control. *J. Fujian Forest. Sci. Technol.* **2015**, *4*, 87–92. (In Chinese)
39. Luo, F.F. The Research about Water Environmental Isotope in Yuanmou Dry-Hot Valleys. Master's Thesis, Yunnan University, Kunming, China, 2012. (In Chinese)
40. Yu, W.J.; Jiao, J.Y. Sustainability of abandoned slopes in the hill and gully Loess Plateau region considering deep soil water. *Sustainability* **2018**, *10*, 1–14. [CrossRef]
41. Li, D.; Fu, Y.P.; Yang, W.Q.; Wu, T.; Xu, L. Effects of different nitrogen and phosphorus levels on the photosynthetic physiology and biomass of *Pinus yunnanensis* seedlings. *J. Anhui Agric. Sci.* **2010**, *38*, 3217–3219. (In Chinese)
42. Yang, W.Y. Community structure and natural regeneration of natural Yunnan pine forest in meddle Yunnan, China. Ph.D. Thesis, Chinese Academy of Forestry, Beijing, China, 2010. (In Chinese)
43. Benítez-Rodríguez, L.; Gamboa-deBuen, A.; Sánchez-Coronado, M.; Alvarado-López, S.; Soriano, D.; Méndez, I.; Vázquez-Santana, S.; Carabias-Lillo, J.; Mendoza, A.; Orozco-Segovia, A. Effects of seed burial on germination, protein mobilisation and seedling survival in *Dodonaea viscosa*. *Plant Biol.* **2014**, *16*, 732–739. [CrossRef] [PubMed]
44. Shanmugavasan, A.; Ramachandran, T. Investigation of the extraction process and phytochemical composition of preparations of *Dodonaea viscosa* (L.) Jacq. *J. Ethnopharmacol.* **2011**, *137*, 1172–1176. [CrossRef]
45. Zhang, Q.; Liu, L.W.; Li, J.Q.; Wang, C.H. Characteristics of *Dodonaea viscosa* population regenerated in abandoned cropland of Yunnan dry-hot valleys, Southwest China. *J. Agric. Sci.* **2016**, *29*, 2234–2238. (In Chinese)
46. Mu, J.; Li, Z.B.; Li, P.; Yu, G.Q.; Zhang, X.X. Applied Effect of Water Retaining Agent upon Vegetation Restoration of Abandoned Dreg Site of Hydropower Station in the Dry-Hot Valley Areas. *J. Xian Univ. Technol.* **2009**, *25*, 151–155. (In Chinese)
47. Wu, J.X.; Chen, Q.B.; Wang, K.Q.; Zhao, Y.Y.; Tong, Z.L. Soil and water conservation benefits and their impacts on soil organic carbon of different vegetation in Central Yunnan Plateau. *J. Northwest A F Univ.* **2015**, *4*, 141–148. (In Chinese)

Article

Assessment of Small-Scale Ecosystem Conservation in the Brazilian Atlantic Forest: A Study from Rio Canoas State Park, Southern Brazil

Manoela Sacchis Lopes [1], Bijeesh Kozhikkodan Veettil [2,3,*] and Dejanira Luderitz Saldanha [1]

1 Programa de Pós-graduação em Sensoriamento Remoto, Centro Estadual de Pesquisas em Sensoriamento Remoto e Meteorologia, Universidade Federal do Rio Grande do Sul (UFRGS), Porto Alegre 91501-970, Brazil; manoelasm@gmail.com (M.S.L.); dejanira.saldanha@ufrgs.br (D.L.S.)

2 Department for Management of Science and Technology Development, Ton Duc Thang University, Ho Chi Minh City 758307, Vietnam

3 Faculty of Environment and Labour Safety, Ton Duc Thang University, Ho Chi Minh City, Vietnam

* Correspondence: bijeesh.veettil@tdtu.edu.vn

Received: 15 April 2019; Accepted: 15 May 2019; Published: 23 May 2019

Abstract: The efficiency of the environmental management of a territory largely depends on previous surveys and systematic studies on the main elements and conditions of the physical environment. We applied remote sensing and digital image processing techniques (Principal Component Analysis and supervised classification) to Landsat imagery for analyzing the spatiotemporal land cover changes occurred in the Rio Canoas State Park in Brazil and its surrounding area from 1990 to 2016. Reforested areas around the park with exotic species is a part of the region's economy and a number of industries depend on it for raw materials. However, it is a matter of concern to avoid contamination with such invasive species, due to the proximity of the Park. From 1990 to 2004, more than 95% of the study area was unchanged and showed minimal distinction in land cover over the 14 years. This was mainly due to the continuous presence of agricultural monocultures around the Park without significant increases (only 3.1% of land cover change during this period). Regarding the interior of the Rio Canoas State Park, from 1990 to 2004, there was no increase in the area of exposed soil. The analysis of the surrounding areas of the park from 2004 to 2016 showed that 5663.78 ha (12.2% of the area) of the land cover has been changed, in most areas, due to reforestation by *Pinus* sp. Notable changes occurred within the park (established in 2004) between 2004 and 2016—there was a partial regeneration of natural species diversity, a small number of invasive species (*Pinus* sp.) and removal of agricultural activities within the park, which contributed a 6.6% (75.45 ha) change in its land cover. We verified that 92.51% (1048.40 ha) of the areas inside the park were unchanged. The results demonstrated that actions were conducted to preserve the natural vegetation cover within the park and to reduce the impacts of anthropogenic activities, including the invasion of exotic species from the surrounding reforested areas into the natural habitat of the park. Given this, our study can aid the environmental management of the Park and its surrounding areas, enabling the monitoring of environmental legislation, the creation of a management plan, and can guide new action plans for the present study area and can be applied to other similar regions.

Keywords: Brazilian Atlantic Forest; ecosystem conservation; Principal Component Analysis; Rio Canoas State Park

---

1. Introduction

Construction of large hydroelectric dams can induce major habitat loss and degradation in the surrounding areas [1,2]. However, deactivated hydropower landscapes have been used worldwide as an area for tourism development, as well as natural conservation units to compensate the lost ecosystems during the construction of hydroelectric reservoirs since the late 19th century [3]. In recent decades, numerous hydropower landscapes around the world became sites for ecosystem conservation, tourism and are defined as protected areas (PAs). Examples for such models for ecosystem conservation can be found in Brazil [4], Canada [5], Costa Rica [6], and many more.

Brazil has the largest PA system in the world (approximately 220 million ha), even though there has been a reduction in the area of PAs since the late 2000s [7]. The Federal Constitution of Brazil expresses in its article 225 the legal and constitutional duty of transmitting the environmental patrimony in the best of conditions to the future and current generations. Likewise, Law 12,651/2012—the Brazilian Forest Code, in its article 1°-A, item I, confirms Brazil's sovereign compromise with the preservation of its forests and other forms of native vegetation, as well as biodiversity, soil, water resources, and integrity of the climatic system. Based on this fact, it is only with preservation practices and environmental control that we will reach a balance, aiming at the reduction of the direct or indirect degradations caused by anthropogenic activities.

The efficient environmental management of a territory largely depends on previous surveys and systematic studies on the main elements and physical conditions. The inappropriate occupation of space and improper use of natural resources can lead to pressure on environmental systems. Among so many laws that guide and foster the environment, we can highlight Law 9985/2000 which institutes the National System of Nature Conservation Units (SNUC). This law establishes the criteria and norms for the creation, implementation, and management of conservation units. From this, we infer that the conservation units (CUs) and their buffer zone are an essential instrument for the protection of biodiversity, natural processes, and environments involved.

Remote sensing, spaceborne data in particular, has been widely used for the dynamic monitoring of land use changes, biodiversity evolution, management of water resources, and the changes on the earth surface in general [8]. Various algorithms for processing remotely sensed data remove the barriers of the human visual system, facilitating effective interpretation of information contained in satellite imagery [9]. In this study, we used satellite data for detecting the land cover changes occurred in a state park located in southern Brazil and its surrounding areas within a 10 km buffer for the period between 1990 and 2016. This study period will allow us to compare the land cover changes occurred before and after the establishment of the park for natural preservation beside a hydroelectric dam in the region. The study area provides an excellent opportunity to understand how the ecosystem loss associated with dam construction can be compensated after the functioning or deactivation of the power plant. We mapped the regeneration of the natural species inside the park and how the establishment of the conservation area reduced the contamination of exotic species into the park while the surrounding areas of the park were cultivated with *Pinus* sp. for industrial raw materials.

2. Characterization of the Study Area

The study site is the Rio Canoas State Park and its surrounding areas, located in the municipality of Campos Novos, in the state Brazilian of Santa Catarina (Figure 1). The Rio Canoas State Park (RCSP) has an area of 1133 hectares and is situated within the Brazilian Atlantic Forest Limit. Its surrounding area, within a buffer zone of 10 km around the park, was also analyzed and encompasses a small portion of the municipalities of Abdon Batista, Anita Garibaldi, and Celso Ramos. The area of the RCSP was acquired by the company Campos Novos Energia S.A–ENERCAN and donated to the state of Santa Catarina to serve as compensation for the environmental loss during the construction of the Campos Novos Hydroelectric Power Plant and the dam.

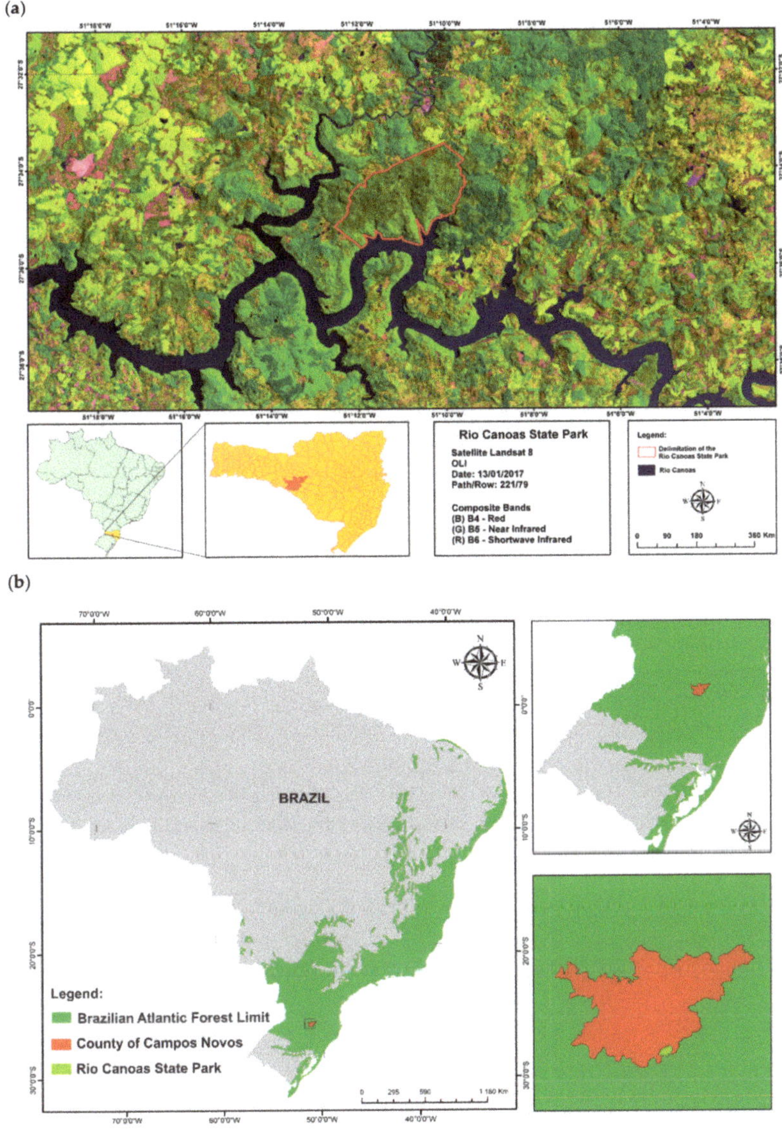

Figure 1. (**a**) Location of the study area in Brazil; (**b**) Location of Rio Canoas State Park (RCSP) within the Brazilian Atlantic Forest Limit.

The study area is characterized by flood basalt, presenting effusive acid rocks on the superior portion and a range of igneous and metamorphic rocks, and the sediments were recently found on the coast [10]. The RCSP is located at the Araucaria Plateau, in the center-west portion of the state. The geomorphology of the study area is, in general, situated on the Geomorphological Unit Rio Iguaçu/Rio Uruguai, characterized by the intense dissection of the plateau along with the main drainage, the Rio Canoas, with large slopes between valleys.

According to the data obtained from the City Hall of Campos Novos, the municipality is known as the breadbasket of Santa Catarina state, with an economy based on the agriculture and is also considered the largest producer of cereals in the state, with the prominent production of corn, soybean, wheat, and beans. The region also focuses on industries, such as cellulose and paper, metallurgy, furniture, and hydropower, which supplies 25% of the state´s consumption. It is worth noting that the raw materials (wood) for the paper and cellulose industries are provided by the pine forests in the surrounding areas of the park. The economy of nearby municipalities (Abdon Batista, Anita Garibaldi, and Celso Ramos) is also based on agriculture, especially corn, bean, soybean, and tobacco crops, and livestock. In Censo Ramos, the cultivation of sugarcane prevails.

According to the phytogeographic data provided by FATMA [11], the original vegetation of this region was represented, primarily, by a Mixed Ombrophilous Forest characterized by the expressive density and physiognomic uniformity of *Araucaria angustifolia* (Brazilian pine), with a sub-grove formed by an expressive number of three species belonging to the Lauraceae family.

3. Datasets

To perform this research, we used satellite imagery, rainfall data and complementary information on the economic and physical characterization of the study area as described below.

3.1. Satellite Data

The list of satellite data is constituted of:

- Three Landsat 8 OLI (Operational Land Imager) images with a 30 m spatial resolution (bands 4, 5, and 6) and band 8 (panchromatic) with spatial resolution of 15 m. Images where acquired on 16 April 2016, 23 September 2016, 9 October 2016 and are available at no cost from the United States Geological Survey (http://www.earthexplorer.usgs.gov).
- Four Landsat 5 TM images with 30 m resolution (bands 3, 4, and 5) and were acquired on 14 July 1990, 16 September 1990, 6 January 1997, 30 March 2004 and 8 October 2004, respectively.
- Global Land Survey (GLS) image of Landsat 5 TM for georeferencing the orbital images of the study area (Source: http://www.dgi.inpe.br/CDSR/).
- WorldView-2 satellite images acquired in 2010, which has a spatial resolution of 0.5 m, provided by FATMA (*Fundação do Meio Ambiente do Estado de Santa Catarina*) and ENERCAN.

3.2. Rainfall Data

The rainfall data near the Campos Novos Hydroelectric Power Plant were acquired from the Hydrological Information System—Hidroweb of the Waters National Agency (http://www.snirh.gov.br/hidroweb/). We used the average of rainfall from 1961 to 1990, historical series of 30 years with monthly information according to the National Meteorology Institute (*Instituto Nacional de Meteorologia*—INMET). Furthermore, we used the total monthly rainfall data obtained from the INMET for the years 1990, 2004 and 2016.

3.3. Complementary Data

In addition to the information mentioned above, we used hydrographic networks of the study area, highway networks, environmental conservation plan in and surrounding area of the dam, the Rio Canoas State Park management plan and information on the agricultural calendar of the main crops cultivated in the state of Santa Catarina, which is provided by the Socioeconomy and Agricultural Planning Center of the State of Santa Catarina (*Empresa de Pesquisa Agropecuária e Extensão Rural de Santa Catarina*–EPAGRI).

3.4. Image Processing System

We used SPRING (*Sistema de Processamento de Informações Georreferenciadas*), an open software package, for processing the images and structuring the databank. We also used ArcGis 10.2 for generating thematic maps and the Microsoft Office Excel 2007 for organizing the graphs and tables, and Garmin™ GPS (Global Positioning System) to collect the field data.

4. Methodology

The methodology adopted to develop this research encompasses various steps as described in the flowchart shown in Figure 2.

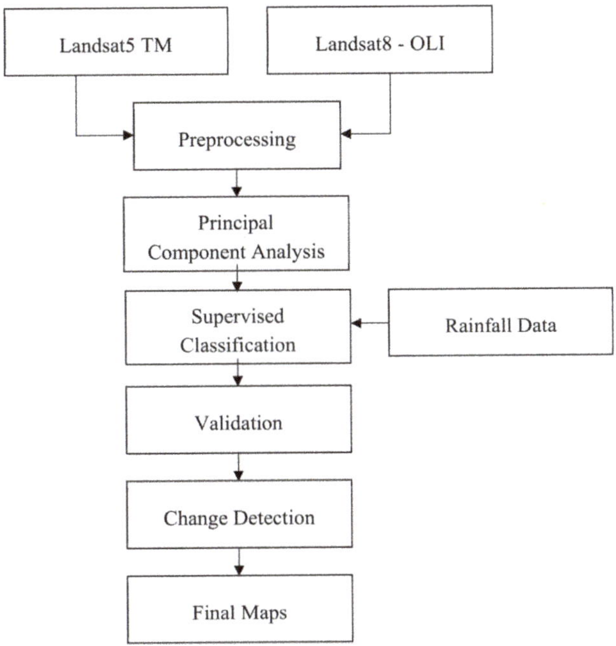

Figure 2. Flowchart of the procedures followed in this study (TM—Thematic Mapper, OLI—Operational Land Imager).

4.1. Image Co-Registration

Image co-registration has been conducted for Landsat *TM* and *OLI* data for eliminating the errors that may occur in the satellite data, due to image rotation, skew and scale. In this case, we proceeded to correct the images, since the study constituted of multitemporal dynamics, in which the images were compared and required to be perfectly coincident in space. Image co-registration is the adjustment of the coordinate system (pixel/lines) of one image to be equivalent to the other image of the same region [9,12]. In this work, we applied pixel resampling method using the nearest neighbor interpolation obtaining the GLS image from the National Institute of Space Research (INPE—*Instituto Nacional de Pesquisas Espaciais*) for choosing the control points. The image co-registration was conducted using SPRING software, in which the first phase was characterized by modifying the extension of the Landsat scenes (path/row: 221/79) for all the years studied with an average of 15 control points, accepting an error of less than one pixel, with a maximum value of 0.55.

4.2. Principal Component Analysis

After the co-registration of images, Principal Component Analysis (PCA) has been conducted with the objective of detecting the land use/land cover (LULC) changes that occurred within the 10 km buffer area surrounding the RCSP. This analysis will allow us to conclude, first, concerning the impact of the construction of the power plant and, second, whether the environmental compensation regarding the Park area was indeed incentivized and implemented properly. Another fact worth mentioning is the inclusion of the buffer zone to analyze is there were any changes in the area surrounding the park for this study to provide subsidy and aid in the future proposal of outlining the buffer zone, given that, until now, it is not present in the management plan or posterior projects. The study of the multitemporal dynamics in 1990, 2004, and 2016 were aided by the Principal Component Analysis to compare the changes before and after the formation of the Campos Novos Hydroelectric Power Plant.

The PCA is a digital image processing technique that uses statistical parameters is considered as efficient in detecting changes in the landscape. In general, the calculation of the principal components of a set of data is conducted by obtaining the eigenvalues and eigenvectors using the correlation matrix or the variance-covariance matrix between the variables of the set [13]. PCA reduces the dimensionality of satellite data that leads to improved data visualization and manageability of data analysis [14,15] and is recognized as one of the best methods for mapping and monitoring interannual and interdecadal vegetation anomalies [16]. The application of PCA, however, depends on the objective of the researcher who can analyze each component according to the work schedule. According to Duarte et al. [17], to identify LULC changes, comparing the differences in the information contained in two or more satellite images is necessary. This identification (of LULC changes) is possible by the application of PCA, which also has the function of determining the extension of the correlation of the image bands and removes them, reducing the dimension of the data and excluding the redundant information that is of no interest (to the user).

Vegetation mapping using automatic methods, such as band rations or vegetation indices are relatively straight forward. However, these methods are not always suitable for differentiating exactly where the changes occurred, but good for what types of changes occurred (native and reforested areas in this case) and PCA has the advantage of mitigating this limitation to some extent by reducing the data dimensionality [14]. Furthermore, the actual nature of changes occurred has been identified by a field visit in this study. It has been observed in a previous study that the overall accuracy and kappa coefficient in vegetation mapping increase when PCA has used along with the main bands of satellite data [18].

The first Principal Component is comprised of the information that is common to all original bands (PC1), the second (PC2) contains the most significant spectral feature of the set. The higher the order of the PCs are, the less significant the spectral features will be. The last principal component has only the information that remained from the set or the noise. We analyzed the correlations of the bands related to the red wavelength (TM3) of the Landsat 5 (1990 and 2004) along with near-infrared (TM4) of the year 2004, generating a new set of images denominated principal components, in a total of three PCs. The use of only three bands is justified by presenting the number of iterations and scenes distinct from when processed by the PCA, resulting in the information of change detection during the period, as confirmed by Lopes [19]. Furthermore, it has been affirmed that the inclusion of the first three PCs corresponds to more than 99.5% of the total variance of satellite data [20].

The same procedure was conducted with the same band combination for comparing the period between 2004 and 2016 after the construction of the RCSP in 2004. The false-color composite image used is comprised of the 2nd (G) and 3rd (R) principal component and the band related to the shortwave infrared (B5) of the original image referent to the most recent year. After applying the PCA, we proceeded to the supervised classification per region based on the Bhattacharyya distance (B). The Bhattacharyya distance can be used as a class separability measure for feature selection [21]. For two normally distributed classes, the Bhattacharyya distance (b) between two classes is defined as:

$$b = \frac{1}{8}(\mu_2 - \mu_1)^{\mathrm{T}}\left[\frac{\Sigma_1 + \Sigma_2}{2}\right]^{-1}(\mu_2 - \mu_1) + \frac{1}{2}\ln\frac{|(\Sigma_1 + \Sigma_2)/2|}{|\Sigma_1|_{1/2}|\Sigma_1|^{1/2}}, \quad (1)$$

where, μi and Σi are the mean vector and covariance matrix of class i, respectively. This algorithm is inbuilt in the SPRING software for image processing.

4.3. Supervised Classification

To proceed with the supervised classification, it is necessary that the user has previous knowledge of the study area. This classification requires the field observation of specific locations shown in the image, from which one can obtain ground-truth data [9,22]. The supervised classification is based on the statistical functions that analyze and compare the characteristics of the spectral reflectance of the pixels associated with a standard class defined by the user. Generally, we calculate the average values, and standard deviations of the defined classes and these values serve as criteria to group the pixels that fulfill the limits close to a specific class [8]. According to Novo [23], the classification process can be distinguished regarding the unit to be grouped. In this work, we used the region-growth algorithm from which we extract homogeneous regions according to the limits tested and established and from the group of contiguous pixels grouped.

The limit of similarity and area used were equal for all images (30). Despite the images being different and from different years, these limits were the most adequate for the identification of LULC. It is worth mentioning that the similarity limit demonstrates the smallest difference accepted between the average value of two pixels (or a set of pixels) and is considered the maximum distance between the spectral centers of two regions. The area limit represents the minimum size of the segment the user wishes to analyze.

In the supervised classification stage, we identified the areas with changes and no changes occurred comparing data acquired between 1990 and 2004, 2004 and 2016, and from 1997 to 2016. Using the false-color composite images from principal components, we observed the areas in which change occurred and those that remained with the same characteristics over the years using the multitemporal Landsat data. We determined the existence of cultivation and reforestation areas with the economic survey and fieldwork studies. However, when analyzing the images, we verified the presence of exposed soil in some regions, with well-formed texture and form. Based on the experience of the user knowledge of the area, allied to the analysis of images from different dates, we perceived that most regions were undergoing an exchange in monocrops or a shallow cutting of exotic species for replanting.

Furthermore, when analyzed the areas in which change occurred or not, we observed an intense increase of the bed of Rio Canoas, due to the creation of the Campos Novos Hydroelectric Power Plant managed by ENERCAN. The power plant is in operation since 2006 and provided approximately $\frac{1}{4}$ of the total energy consumption for the state of Santa Catarina. Therefore, we conducted a supervised classification of the areas in which change occurred on the riverbed before and after the creation of the power plant using a historical raifall data series encompassing 30 years in the Campos Novos station and the monthly total rainfall data for 2016 and 2004 to remove the influence of rainfall on the increase of the riverbed and analyze these differences.

4.4. Fieldwork

The fieldwork was conducted in two stages: From 26 November 2015 to 30 November 2015 and from 3 February 2017 to 4 February 2017 within the RCSP and its surrounding area aiming at a future elaboration of a proposal for buffer zones. The sampling points were selected using the Random Points tool of ArcGis® 10.2 and posteriorly guided according to the need for validating the multitemporal dynamics to identify the land cover at the locations presenting or not changes, as demonstrated on the map elaborated using the PCA. A total of 107 points were analyzed. The tools used in this work are Garmin® Etrex30 GPS, a digital camera (Sony Cyber Shot DSC H300), a clipboard for notes, and maps elaborated for the study area. To identify the areas with the need for change detection in the RCSP, we elaborated a spreadsheet with the features acquired from Landsat OLI images and compared the information with the photographs acquired at the location.

5. Results and Discussion

5.1. Principal Component Analysis

The Principal Component Analysis reduced the redundancy of the information between the spectral bands, which presented very similar behavior. The number of principal components is equal to the number of bands in which each component is associated with a variance of the digital levels, with the first component presenting the highest variance, successively decreasing the values [17,24]. Thus, the application of this technique, manipulated with multiple band association tests, can demonstrate the areas of use dynamic and land cover from 1990 to 2004 and 2004 to 2016.

Similar to the observations made by Ding et al. [20], the first three PCs of each year corresponds to more than 99.5% of the total covariance. For example, for the Landsat TM image in 2004, PC1 presented 68.6% of the total covariance of the set and PC2 and PC3 presented 20.5% and 10.7%, respectively (total 99.9%). Based on this information, we perceived that the second and third components are not correlated. This analysis associated with RGB combination tests between the bands allowed the identification of the changes that occurred in 1990 and 2004.

After the parameter analysis, we proceeded to study the combination of the principal components and the bands on the RGB composition. The composition that demonstrated the areas in which changes occurred or not was the second component regarding the green channel (G), the third component regarding the red channel (R), and band 5 (intermediate infrared–L5–2004) regarding the blue channel (B), as demonstrated in Figure 3a. This composition of bands best met the objectives after numerous tests for detecting changes in the reforestation areas from 1990 to 2004 and from 2004 to 2016.

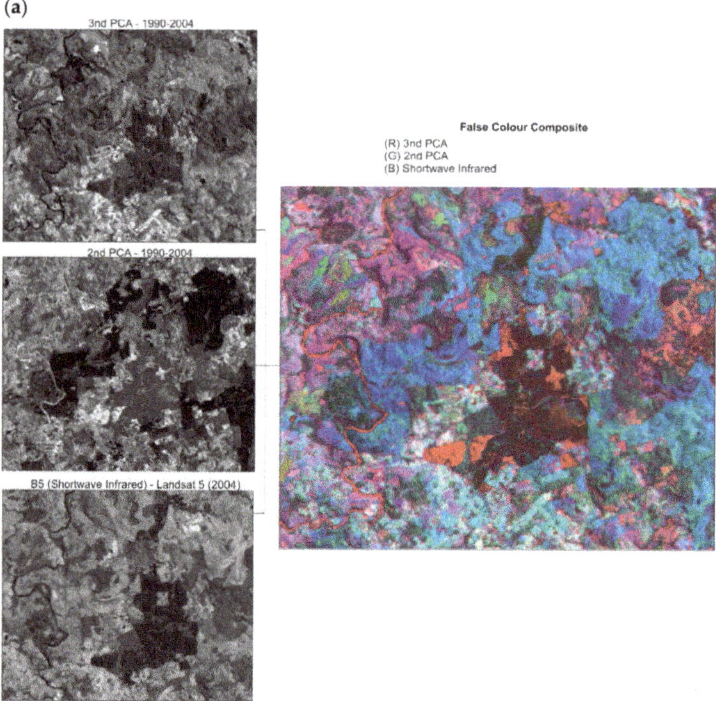

Figure 3. *Cont.*

Figure 3. (**a**) False-colored composites using 2nd and 3rd Principal Components and Band 5 of the Landsat TM (2004); (**b**) Change detection in the reforested areas (2004 to 2016).

From Figure 3a, we can analyze that, the areas in dark green as LULC that had already existed in 1990 and continued in 2004. The areas colored in light and dark orange demonstrated the changes that occurred in the last 14 years. The areas in dark blue are the regions that existed in 1990 and showed changes in 2004. This combination indicated very effectively the respective changes/no-changes, as shown in Figure 3b, which established the comparison between the images acquired in 2004 and 2016.

It is worth mentioning that the result of the PCA highlighted the information measured and the changes that occurred between years (of image acquisition). However, if in an image of a specific date the soil was exposed, and, in another, the areas were cultivated, the color composition image will consequently present an area of land use and cover change. At this moment, the analyst must add the ground-truth data from the study area obtained during the fieldwork, images from other dates, which corroborated with the identification of areas with temporary crops and reforestation areas distinct from those used for the classification.

5.2. Supervised Classification

The supervised classification technique was used to elaborate on the LULC change detection map of the RCSP and its surrounding area. In this work, we applied the supervised classification based on the Bhattacharyya distance measurement. This classification divided the homogeneous regions of the images according to the area and similarity limits indicated. As Figure 4a shows, we used an area and similarity limit equal to 30 for both images originated from PCA.

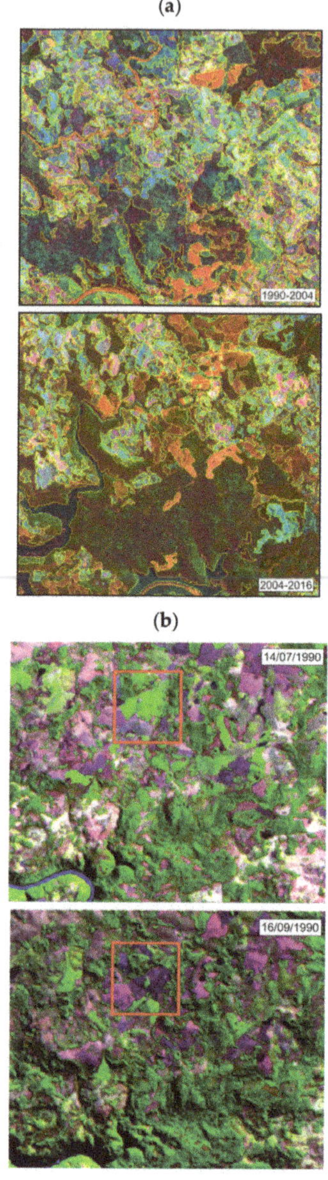

Figure 4. (**a**) Segmentation of the homogeneous areas of the images (limit of the area and similarity equal to 30) (**b**) Exemplification of crops in less than two months. Difference between the exposed soil and crop exchange.

The training samples were indicated after a previous study of the area and its LULC using the high-resolution image (WorldView-2) provided by ENERCAN, images from different dates to obtain a better measure, linear contrast and segmentation techniques, and fieldwork. Moreover, to aid in the classification of the areas that presented changes, but that, for their texture and well-defined form, were similar to temporary crops, we abandoned the use of the agricultural calendar from the Socioeconomic and Agricultural Planning Center (CEPA) from EPAGRI.

The data from the agricultural calendar are presented by micro-regions, encompassing the municipalities in the areas surrounding the Rio Canoas State Park. The study area presented higher representativeness of corn, soybean, tobacco, bean, and wheat crops. According to CEPA, in 2016, 75% of the corn crops and 95% of tobacco were planted in October, while the 85% of the soybean was planted between November and December, and 60% of the bean crop, in December. The winter crops, such as wheat, were planted in June, July, and August. This information associated with the orbital images from different dates was essential concerning the supervised classification because, depending on the date of the image, the cultivation areas could present exposed soil, indicating the times of production rest, areas improper for cultivation or the moment between crops.

Therefore, with the objective of more accurately classifying the agricultural areas and exposed soil, we used the images from different dates to befit the times of production for most crops in the region. We discriminated five categories established on the theme caption, change and no-change, monocrops, water bodies, and changes in water bodies. The monocrop class encompasses the areas with agriculture crops, reforestation, and pasture, since they often present a similar spectral behavior and distinction, not as an object of this work, but presenting or not changes in the land cover. The training and test samples were acquired in an average of 100 for the classification of the false-color composites of the principal components associated with the original band of the intermediate infrared. The PCA allowed the identification of changes and no-changes in the same image, also reducing the classification time of the area, and highlighting the distinct theme classes. After classifying the images, we conducted the post-classification procedures and matrix edition of the classes presented as changed areas with exposed soil, but that, according to many studies and analyses, we concluded that these areas were recently harvested or planted agriculture areas with no plant growth, as shown in Figure 4b.

After the classification of the image, validation has been performed visually and mathematically using the control points collected during the two field trips. More than 100 random points distributed in the image were analyzed using ground truth data from the study area with the aid of GPS. When considering all land cover classes defined, the overall accuracy index was 0.93 and the Kappa index was 0.88. This value of Kappa is associated with the quality of the classification is considered as good according to Landis and Koch [25].

5.2.1. Change Detection within the Rio Canoas State Park

The Rio Canoas State Park is a conservation unit of a mixed ombrophilous forest or araucaria forest with approximately 1200 hectares. Because of this, previous to 2004, the RCSP had not yet been created and, consequently, preservation was not required. The term Conservation Unit (CU) is defined by the MMA [26] as the territorial space and its environmental resources, including the jurisdictional waters, with relevant natural characteristics, legally instituted by the Public Power enterprise, with objectives of conservation and defined limits, under special administration regime, to which adequate protection guarantees are applied.

According to the Law nº 9985 of 18th July 2000, the conservation units integrating the National Nature Conservation Units System are divided into two groups with specific characteristics: The integral protection units and the sustainable use units. The objective of the integral protection units is to preserve nature and admits only the indirect use of its natural resources, except in certain legal cases. The general objective of the sustainable use units is to harmonize environmental conservation with the sustainable use of a portion of its natural resources.

Among the integral protection groups, the units considered are the Ecological Station, Biological Reserve, National Park, Natural Monument, and Wild Life Refuge. The National Park, the object of this work, has a specific objective of the preservation of natural ecosystems of high ecological relevance and scenic beauty, allowing the performance of scientific researches and development of environmental education activities, recreation in contact with nature, and ecological tourism. Based on this, we classified the images resultant from the PCA, comparing the periods between 1990 and 2004 and between 2004 and 2016, as presented in Figure 5a,b.

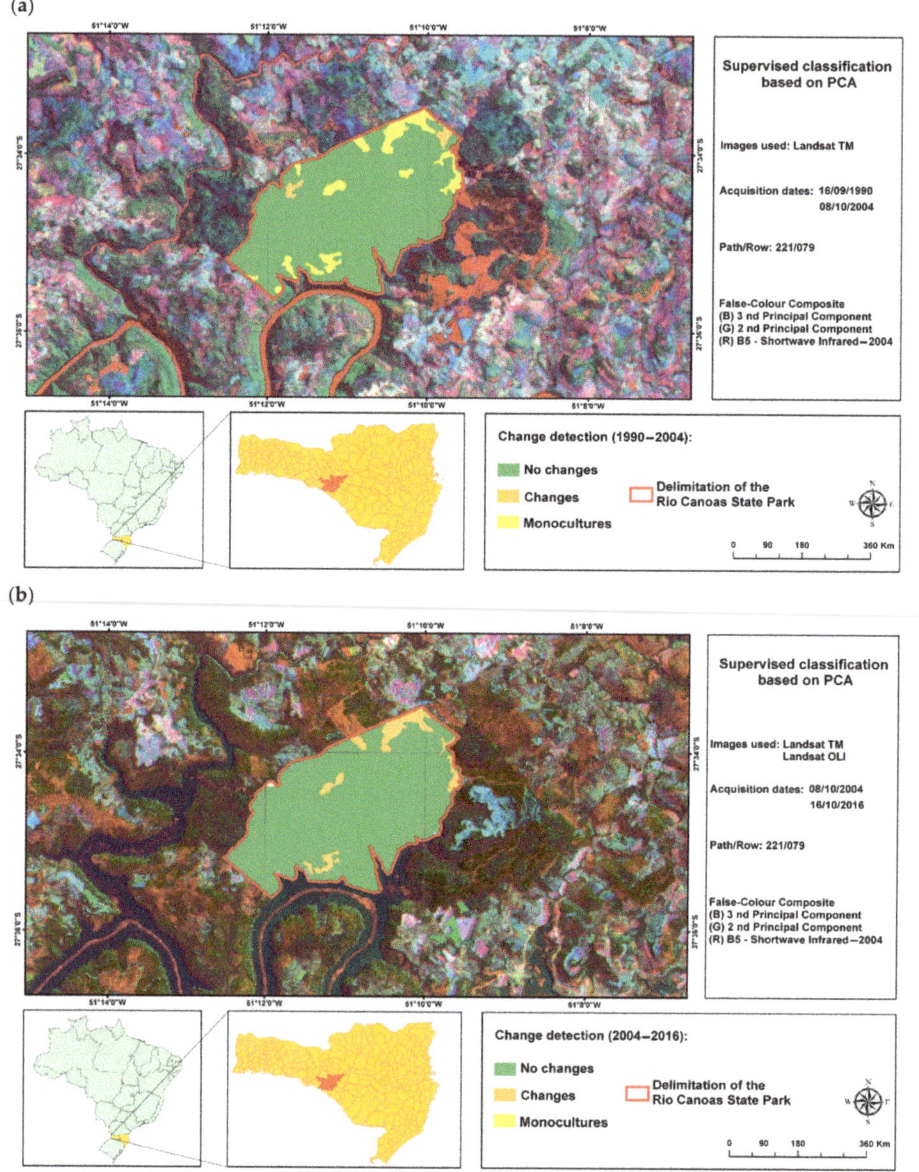

Figure 5. (**a**) Change detection map for Rio Canoas State Park (1990–2004); (**b**) Change detection map for Rio Canoas State Park (2004–2016).

Before the creation of the RCSP up until a few months after the Decree was approved, we verified that there were crops and plantations of exotic species within the park in areas significant to the conservation unit. However, with the consolidation of the park and over time, its interior was modified.

In Figure 4b, we observed the presence of monocrop class which encompasses areas of agricultural and silvicultural activities, such as soybean, corn, tobacco, and exotic species of the Pinus and Eucalyptus genre. In the change detection map referent to the comparison from 1990 to 2004, we verified that the areas with cultivations within the RCSP remained from 1990 until the date in which the image from 2004 (10 August 2004) was taken. These areas presented changes when comparing the images using the PCA. However, by studying other images, we verified that the area was undergoing crop exchange. Associated to this, the class measures from the SPRING software showed that 89.56% (1014.94 ha) were unchanged areas and 10.44% (118.32 ha) were areas with monocrops. Thus, we verified that, during the period from 1990 to 2004, there were no increments to the area of exposed soil, but changes of agricultural and forestry crops. Figure 5a shows an obvious distinction between the years of 2004 and 2016. We observed that 92.51% (1048.40 ha) are unchanged areas, 6.6% (75.45 ha) demonstrated changes, and 0.7% (7.92 ha) is the riverbed that entered the Park with the creation of the Campos Novos Hydroelectric Power Plant.

The multitemporal dynamic of the land cover of 2004 and 2016, aided with the fieldwork, images from other dates, and the experience of the managers of the conservation unit indicated that, over the years, the Park aimed for changing most monocrops in significant areas to regenerate other forestry species, which ratifies the objective of the conservation unit. As demonstrated in Figure 5b, we verified that two of the areas which the image-product classification of the PCA indicated as changed areas were real. These were areas previously occupied by Pinus species and removed to regenerate species native of the region.

5.2.2. Change Detection in the Surrounding Areas of RCSP

We analyzed the 10 km buffer area surrounding the RCSP to verify the vegetation coverage in the area after the formation of the Campos Novos Hydroelectric Power Plant. This verification allowed us to raise conclusion, first, on the impact the construction of the power plant and, second, if the environmental compensation regarding the PAs of the Park has been implemented correctly. Another fact concerns the inclusion of a Buffer Zone to analyze whether many changes occurred in the area surrounding the Park for this study to provide a subsidy and aid in the posterior proposal of a Buffer Zone, since it briefly mentioned in the management plan. According to the data presented in Table 1 and the same steps are taken to detect changes within the Park, the areas presenting the most changes in land cover were those of the images taken in 2004 and 2016.

Table 1. Classification table for change detection for the area surrounding the RCSP.

	Changes from 1990 to 2004	
Classes	**Area (ha)**	**Area (%)**
Change	1436.56	3.10%
No-change	18,477.92	39.81
Monocrops No change	25,738.98	55.46%
Rio Canoas	760.34	1.64%
Total	46,413.8 ha	100%
	Changes from 2004 to 2016	
Classes	**Area (ha)**	**Area (%)**
Change	5663.78	12.20%
No-change	21,862.20	47.10%
Monocrops No-change	16,151.54	34.80%
Rio Canoas	796.04	1.72%
Increase of the Rio Canoas	1939.86	4.18%
Total	46,413.8 ha	100%
2016–Global Accuracy: 0.93		
2016–Kappa: 0.88		

During the period from 2004 to 2016, the land cover changes in 5663. 78 ha (12.20% of the area), mostly by reforestation of Pinus species (Figure 6b). The surrounding area of the park is characterized by cellulose and paper industries, such as *Iguaçu Celulose S.A*, and wood industries. The reforestation areas of exotic species surrounding the RCSP is old and a part of the region´s economy (Figure 6c,d).

However, this is a concerning subject for avoiding contamination with invasive species, due to the proximity to the park.

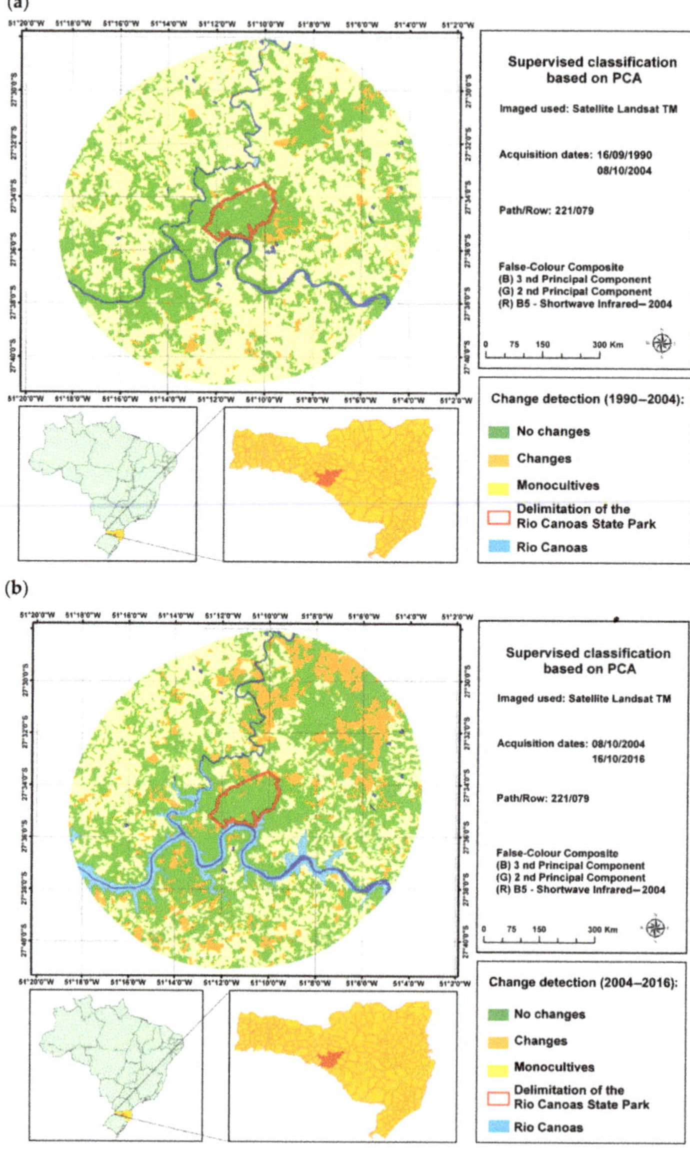

Figure 6. *Cont.*

Figure 6. (**a**) Change detection map for the period from 1990 to 2004; (**b**) Change detection map for the period from 2004 to 2016; (**c**) Worldview-2 image acquired in 2010 showing the areas of regenerated vegetation (Courtesy: ENERCAN); (**d**) Photograph of native species *Araucaria angustifolia* with the plantations of *Pinus eliotti*; (**e**) Photograph of various native species and *Pinus eliotti*.

In 1990 and 2004, the areas that remained unchanged were of approximately 95% of the study area and presented no changes in land cover during the 14 years (Figure 6a). This occurs mainly due to the continuous presence of agricultural monocrops in both summer and winter, with no great increments. Regarding the 12 years from 2004 to 2016, the changes were more recurrent. This occurs mostly due to the shallow cutting of the Pinus plantation areas, which differ from the crops, such as soybean, tobacco, and corn, and changes the entire landscape. The cutting of the exotic reforestation species take years and is performed in rotations that ranges from 5 to 8 years using shallow or partial cuts. Therefore, in these cases, there is change and, contrary to the soybean crops, for example, which can be replanted as soon as the wheat is harvested and germinates days later, the plantation of new seedlings takes time.

5.2.3. Change Detection in the Bed of the Rio Canoas

The Campos Novos Hydroelectric Power Plant was implanted on the Rio Canoas, approximately 20 km upstream of its confluence with Rio Pelotas at the border of the municipalities of Campos Novos and Celso Ramos, according to the data provided by ENERCAN [27]. According to ENERCAN, before finishing the construction at the margins of the Rio Canoas, the dam would be formed by the flooding of the peripheral areas to the Rio Canoas, characterizing a substantial extension of flooded land.

The PCA has demonstrated the transformations as pursuant to the study of the RCSP associated with the change detection. Because the data on the flooding of the marginal areas provided by the ENERCAN Company were provided before concluding the construction in 2004, we proceeded with the analyses of the area. Therefore, the data were classified in two images with specific dates and related to the rainfall data, derived from Landsat TM (acquired on 14 July 1990) and Landsat OLI (acquired on 16 April 2016).

To obtain a better accuracy of the image classification, we acquired the rainfall data from the Hydrological Information System—Hidroweb of the National Waters Agency (Figure 7). We used the average rainfall from 1961 to 1990, historical series of 30 years, with monthly information according to the National Meteorology Institute (INMET). To classify the images, they were selected according to the dates in which there was monthly accumulated rainfall nearly the same as the average for the 30 years analyzed to exclude the influences of rain in the demarcation of the riverbed before and after flooding the area.

As presented in Figure 7, the areas in light pink identify the dates with rainfall accumulations that could influence the increase of the riverbed. We used the images according to the availability and the value nearest to the average of rain accumulated in 30 years. The image acquired on 6 January 1997 demonstrates that the amount of rain accumulated was within the average and, therefore, there were not many influences of rain. Regarding the image acquired on 16 April 2016, when considering rainfall data in October, we verified a significant increase of accumulated rainfall, but we disregarded the daily averages of accumulated rainfall for October of 2016. With this detailed information from INMET, we verified that, until the 16th of October, the rainfall accumulation was of 78 mm, also having little effect in the area. According to the generated map (Figure 8), an area of 6.48 km^2 has been indicated as an approximate measure of the riverbed on 6 January 1997 and of 30.15 km^2 on 16 April 2016 with the creation of the Campos Novos Hydroelectric Power Plant. The objective of the study was to analyze the size of the area surrounding the RCSP that was flooded, due to the high difference made explicit in the change detection using the PCA technique.

Figure 7. Rainfall accumulated at the time of satellite data acquisition compared to the average of 30 years.

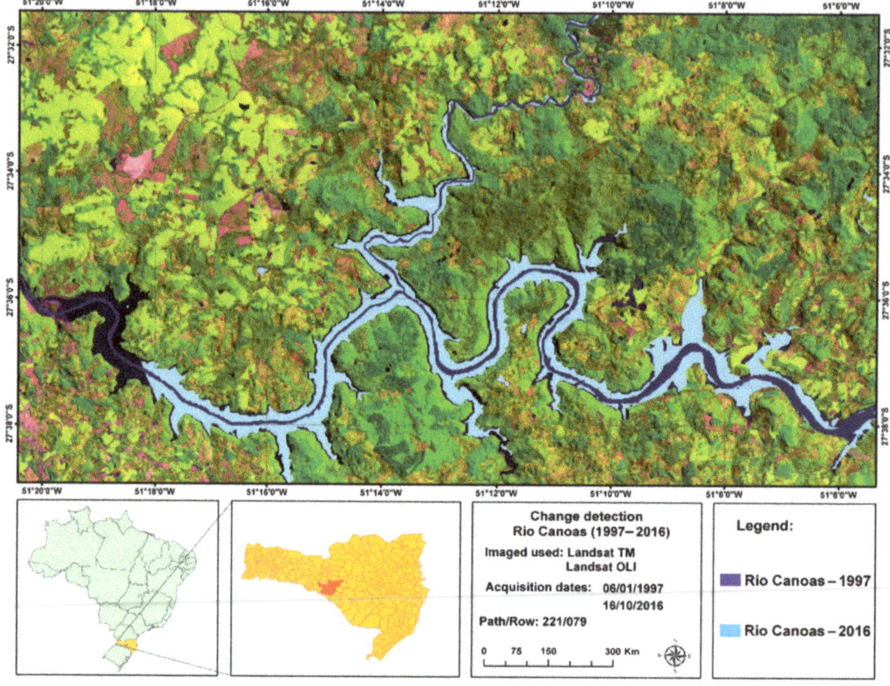

Figure 8. Map of the Rio Canoas in 1997 before the implementation of the Hydroelectric Plant of Rio Canoas, and in 2016.

6. Conclusions

The Rio Canoas State Park is the largest remaining portion of the Mixed Ombrophilous Forest (Araucaria Forest) surrounding the Campos Novos Hydroelectric Power Plant dam, in the southern region of Brazil. The forest ecosystem belongs to the Atlantic Forest Domain of South America, which is under the threat of extinction in the Santa Catarina state of Brazil.

Based on the work conducted, we conclude that the Principal Component Analysis associated with specific bands after conducting a field study is effective in detecting LULC changes when comparing two or more dates, particularly in understanding exactly where the changes occurred rather than what changes occurred.

The analysis conducted in the area surrounding the Rio Canoas State Park, which is established in 2004, for the period between 1990 and 2004 demonstrated that more than 95% of the park and the surrounding areas remained intact. The area, where the park is situated had no land cover change during this period and 3.1% of the surrounding areas showed changes, due to agriculture and silviculture practices. This change was minimal due to the continuous presence of agricultural monocrops in both summer and winter in the area surrounding the Park.

For the period between 2004 and 2016, it is observed that about 5663.78 ha (12.20% of the area) of the land cover surrounding the park changed, due to the reforestation of Pinus species. Within the park, we verified that 92.51% (1048.40 ha) are no-change areas, 6.6% (75.45 ha) presented changes (this includes regeneration of native species, land cover changes, due to the abandoning of agriculture within the park and invasive species), and the rest (0.89%) is the riverbed that entered the Park after the creation of the Campos Novos Hydroelectric Power Plant. This data indicates actions regarding the preservation of the vegetation cover within the Park to reduce the impacts of anthropic activities.

The areas surrounding the Park are characterized by cellulose, paper and wood industries. The reforestation areas with exotic species (e.g., *Pinus elliottii*, *P. taeda*) surrounding the Park are old and a part of the region´s economy. From an ecological point of view, it is important to avoid contamination of natural forest in the protected areas by such invasive species (*Pinus* sp. in particular), due to the proximity to the Park.

Furthermore, we demonstrated an increase in the Rio Canoas bed with the implementation of the Campos Novos Power Plant and concluded that there was an increase of 23.67 km^2 in 2016 when compared to 1997. This area advanced the initial limits of the Rio Canoas State Park, changed the landscape, caused the translocation of people, and flooding of the areas previously covered by ciliary forests. The results of this study can be used in the environmental management of similar kinds of parks and surrounding areas, allowing the monitoring of the environmental legislation and management plan, and guide plans of action.

In a nutshell, it was concluded that after the creation of the Rio Canoas State Park, some modifications were made to its LULC, such as the clearing of larger areas with Pinus plantations, in order to facilitate the regeneration of native species and to create projects to control the expansion of exotic species. The large-scale plantations of exotic species exist prior to the creation of the park and these areas contribute to the economy of the region. However, it is known that reforestation with these invasive species, such as *Pinus eliotti* and *P. taeda*, depending on how the plantations are conducted, may lead to the decrease in some of the native species and this may alter the natural diversity of the ecosystem. After the creation of the park, expansion of exotic species into the park has been reduced, due to the control measurements taken by the authorities. These factors project the importance of a time-series monitoring of the forest fragments of mixed ombrophilous forests belonging to the core zone of the Brazilian Atlantic Forest Biosphere Reserve.

Author Contributions: All authors contributed equally to the text; M.S.L. and D.L.S. conducted field studies and B.K.V. prepared the figures.

Funding: This research received no external funding.

Conflicts of Interest: The authors declare no conflict of interest.

References

1. Palmeirim, A.F.; Peres, C.A. Rosas FCW Giant otter population responses to habitat expansion and degradation induced by a mega hydroelectric dam. *Biol. Conserv.* **2014**, *174*, 30–38. [CrossRef]
2. Lees, A.C.; Peres, C.A.; Fearnside, P.M.; Schneider, M.; Zuanon, J.A.S. Hydropower and the future of Amazonian biodiversity. *Biodivers. Conserv.* **2016**, *25*, 451–466. [CrossRef]
3. Rodriguez, J.F. Hydropower landscapes and tourism development in the Pyrenees. *J. Alp. Res.* **2012**. [CrossRef]
4. Rylands, A.B.; Brandon, K. Brazilian Protected Areas. *Conserv. Biol.* **2005**, *19*, 612–618. [CrossRef]
5. Sherren, K.; Beckley, T.M.; Parkins, J.R.; Stedman, R.C.; Kelity, K.; Morin, I. Learning (or living) to love the landscapes of hydroelectricity in Canada: Eliciting local perspectives on the Mactaquac Dam via headpond boat tours. *Energy Res. Soc. Sci.* **2016**, *14*, 102–110. [CrossRef]
6. Bernard, F.; de Groot, R.S.; Campos, J.J. Valuation of tropical forest services and mechanisms to finance their conservation and sustainable use: A case study of Tapantí National Park, Costa Rica. *For. Policy Econ.* **2009**, *11*, 174–183. [CrossRef]
7. Bernardi, D.; Disperati, A.A.; Santos, J.R.; Mendes, F.S. Monitoramento da dinâmica de paisagem através da análise por componentes principais (ACP) em imagens Landsat 5. *Simpósio Bras. Sens. Remoto* **2001**, *10*, 557–560.
8. Liu, W.T.H. *Aplicações de Sensoriamento Remoto*; Uniderp: Mato Grosso do Sul, Brazil, 2006; 881p.
9. Crósta, A.P. *Processamento Digital de Imagens de Sensoriamento Remoto*; UNICAMP: Campinas, Brazil, 1992; 170p.
10. Duzzioni, R.I. *Elaboração dos Planos de Manejo do Parque Nacional das Araucárias e da Estação Ecológica da Mata Preta*; Icmbio: Brasília, Brazil, 2005; 150p.

11. Fundação do meio ambiente de Santa Catarina (FATMA). Parque Estad ual das Araucárias–Plano de Manejo/Diagnósticos. Florianópolis-Santa Catarina. 2007. Available online: http://www.fatma.sc.gov.br/upload/ucs/araucarias/Plano_Manejo_PAEAR_Fase_II_final.pdf (accessed on 18 September 2018).
12. Vittek, M.; Brink, A.; Donnay, F.; Simonetti, D.; Desclée, B. Land Cover Change Monitoring Using Landsat MSS/TM Satellite Image Data over West Africa between 1975 and 1990. *Remote Sens.* **2014**, *6*, 658–676. [CrossRef]
13. Coutinho, M.D.L.; Brito, I.B. Análise de componentes principais com dados pluviométricos no estado do Ceará. In Proceedings of the Congresso Amazônia e o clima global, Belém-Pará, Brazil, 13 April 2010; Volume 1.
14. Munyati, C. Use of Principal Component Analysis (PCA) of remote sensing images in wetland change detection on the Kafue Flats, Zambia. *Geocarto Int.* **2004**, *19*, 11–22. [CrossRef]
15. Lattin, J.M.; Carrol, J.D.; Green, P.E. *Analyzing Multivariate Data*; Thomson Brooks/Cole: Singapore, 2004; p. 556.
16. Lasaponara, R. On the use of Principal Component Analysis (PCA) for evaluating interannual vegetation anomalies from SPOT/VEGETATION NDVI temporal series. *Ecol. Model.* **2006**, *194*, 429–434. [CrossRef]
17. Duarte, C.C.; Souza, S.F.; Galvíncio, J.D.; Melo, I.D.F. *Detecção de Mudanças na Cobertura Vegetal da Bacia Hidrográfica do Rio Tapacurá-PE Através da Análise por Componentes Principais*; Simpósio Brasileiro de Sensoriamento Remoto: São José dos Campos, Brazil, 2009; pp. 5765–5772.
18. Khedmatgozar, M.; EslamBonyad, A. Use of Principal Component Analysis in accuracy of classification maps (case study: North of Iran). *Res. J. For.* **2016**, *10*, 23–29. [CrossRef]
19. Lopes, M.S. Análise de vulnerabilidade natural à erosão como subsídio ao planejamento ambiental do oeste da bacia hidrográfica do Camaquã–RS. *Rev. Bras. Cartogr.* **2016**, *68*, 1689–1708.
20. Ding, J.L.; Wu, M.C.; Tiyip, T. Study on soil salinization information in arid region using remote sensing technique. *Agric. Sci. China* **2011**, *10*, 404–411. [CrossRef]
21. Choi, E.; Lee, C. Feature extraction based on the Bhattacharyya distance. *Pattern Recognit.* **2003**, *36*, 1703–1709. [CrossRef]
22. Guerschman, J.P.; Paruelo, J.M.; Bella, C.D.; Giallorenzi, M.C.; Pacin, F. Land cover classification in the Argentine Pampas using multi-temporal Landsat TM data. *Int. J. Remote Sens.* **2003**, *24*, 3381–3402. [CrossRef]
23. Novo, E.M.L. *Sensoriamento Remoto Princípios e Aplicações*; Edgard Blücher: São Paulo, Brazil, 2010; 388p.
24. Jensen, J.R. *Introductory Digital Image Processing*; Prentice Hall: Upper Saddle River, NJ, USA, 2005; 544p.
25. Landis, J.R.; Koch, G.G. The measurement of observer agreement for categorical data. *Biometrics* **1977**, *33*, 159–174. [CrossRef] [PubMed]
26. Ministério do Meio Ambiente (MMA). Plano Estratégico Nacional de Áreas Protegidas–PNAP. Brasília. 2006; 44p. Available online: http://www.mma.gov.br/estruturas/205/_arquivos/planonacionaareasprotegidas_205.pdf (accessed on 18 September 2018).
27. ENERCAN. *Campos Novos Energia S.A*; Plano de Manejo da Usina Hidrelétrica Campos Novos: Campos Novos, Santa Catarina, Brazil, 2004; 25p.

Article

Regeneration Ecology of the Rare Plant Species *Verbascum dingleri*: Implications for Species Conservation

Petros Ganatsas [1,*], Marianthi Tsakaldimi [1], Christos Damianidis [1], Anastasia Stefanaki [2], Theodoros Kalapothareas [1], Theodoros Karydopoulos [1] and Kelly Papapavlou [3]

1 Laboratory of Silviculture, Department of Forestry and Natural Environment, Aristotle University of Thessaloniki, P.O. Box 262, University Campus, GR54124 Thessaloniki, Greece; marian@for.auth.gr (M.T.); cdamiani@for.auth.gr (C.D.); tkarydop@for.auth.gr (T.K.); theodorekal@gmail.com (T.K.)

2 Naturalis Biodiversity Centre, 2332 AA Leiden, The Netherlands; anastasia.stefanaki@gmail.com

3 EXERGIA S.A. Voukourestiou 15, GR-106 72 Athens, Greece; kpapapavlou@ath.forthnet.gr

* Correspondence: pgana@for.auth.gr

Received: 8 March 2019; Accepted: 4 June 2019; Published: 15 June 2019

Abstract: *Verbascum dingleri* Mattf and Stef. is a Greek endemic plant species belonging to the family of Scrophulariaceae that only occurs in northeastern Greece, east of the city of Kavala. Knowledge of species distribution, habitat requirements, reproduction, ecology, and population characteristics is limited in the literature. In this study, habitat characteristics, population counts, fruit and seed diversity, and germination were studied for the first time. The results indicate that the species geographical distribution is very restricted, lying in the Mediterranean floristic zone at a low altitude (100–200 m asl) and on very shallow soils. The habitat of this species is characterized by the Csa climate type, with a mean annual precipitation of 602 mm and a mean annual temperature of 14.6 °C. The species occurs in the area lying between the geographical coordinates 40°58′16.59″ N, 24°27′54.93 E, and 41°05′7.2″ N, 24° 47′17.2″ E. The species thrives in degraded shrub communities, dominated by the shrub species *Paliurus spina-cristi* Mill., *Olea europea* L. ssp. *europaea*, and *Quercus coccifera* L. Only a very small number of individuals were found (less than 200) at a density considered too small for long-term persistence of the species. The fruits of the species contained a high number (mean value 58.2) of minute seeds. The seeds exhibited high germination (up to 80.0% in laboratory and up to 30% in ambient conditions). We conclude that in situ and ex situ species conservation and habitat restoration are feasible through the introduction of seedlings produced from seeds collected from local populations.

Keywords: endangered plant; ex situ conservation; plant reintroduction; seed germination; seedling propagation

---

1. Introduction

Over the last decades, many plant species have been threatened as a consequence of several anthropogenic disturbances and climate change that resulted in species habitat loss. Especially, species with very limited distribution ranges, specific site requirements, and small populations are at high risk of extinction [1].

Unfavorable site conditions for species self-reproduction and seedling development in natural or human-disturbed habitats may cause difficulties in species ecological status and future prospects [2–4]. Thus, data on seed ecology and species regeneration requirements are of great importance.

Early phases of plant life such as germination, seedling development, and early growth are very important stages in plant life cycle and greatly vary in space and time, especially for species of limited populations [5,6]. These stages are commonly affected by many endogenous (lethargy or desiccation tolerance) and exogenous (e.g., site conditions) parameters that greatly threaten species life [7,8]. Successful seed germination and seedling propagation are key factors for ex situ species conservation and in situ species restoration [9,10].

Verbascum dingleri Mattf and Stef. is a rare Greek plant species that belongs to the family Scrophulariaceae, with extremely narrow geographical distribution in northern Greece [11–14]. According to the available information, its presence worldwide has been recorded only three times [12] in two localities, as reported in the Global Biodiversity Information Facility (GBIF) Backbone Taxonomy Checklist Dataset (https://doi.org/10.15468/39omeiassessed via GBIF.org): one time, near Chrysoupolis in 1936, in a location with coordinates 40,98 N, 24,70 E, at an altitude of 150 m above sea level, and two times, in Toxotes, close to the first record (at a distance of 15 km) and at the same elevation, on limestone. The first species record was described by Mattfeld in 1926 [11]. Very few further records have been found in the Greek "grey" literature, but these were omitted in the manuscript. The high frequency of anthropogenic disturbances in the area where the species lives seems to affect the species conservation status. The target species population was located adjacent to a linear construction project which had commitments to mitigate biodiversity impacts. The commitments included seed collection for post construction habitat restoration initiatives.

In this study, the autecology of *V. dingleri* was investigated to provide a better understanding of the conservation of this species. The research was focused on determining the species distribution range and habitat area, community characteristics, population information, and reproduction abilities for in situ and ex situ conservation. More specifically, the study aimed at:

(i) Determining the environmental drivers forcing species distribution;
(ii) Studying species fruit and seed diversity;
(iii) Investigating seed germination behavior in laboratory (controlled environment) and ambient conditions (nursery), including the effect of fruit characteristics on seed germination;
(iv) Developing a scientific approach for the species and its habitat conservation by producing seedlings from seeds in a nursery.

2. Materials and Methods

2.1. Species Description

V. dingleri is [11] a perennial herb species that grows up to approximately 60 cm. The stem is erect, slender, with parallel lines or grooves. The inferior part of the stem is greyish with stellate hairs, without glands, and the upper part is smooth and without hairs. The basal leaves (rosette) are lax with stellate hairs on both sides, denser on the lower side. The shape of the basal leaves is oblong, inversely ovate, the inferior part pinnate, with a length of approximately 12 cm and a width of 3–5 cm. Basal leaves with peduncles have a length of about 3 cm and a width of about 0.5 cm. Leaves' base is flat above and convex underneath. The wings on either side of the petiole inflorescence with simple or complex panicles more or less branched, open, erect and slender, 15–30 cm long, racemose. The flowers with peduncles can be solitary, 5–7 mm long, with a little bract that appears ovate-lanceolate, acute, without hairs. The calyx is more or less conic, with a length of 3–3.5 mm and a diameter of 2.5 mm, without hairs. The calyx lobes are linear-lanceolate, acute or sub-obtuse, with a length of 2.8–3 mm and a width of 0.8–1 mm, and present margins with minutely sparse glands, with obscure three-nerved and middle-nerve keel. The corolla is yellow, about 1.5–2 cm in diameter, without hairs on both sides; the corolla tube is about 3 mm long, with rounded inversely ovate lobes. The anthers are reniform, about 1.5–1.8 mm long, the stamens filament are about 3 mm long and 2.5 mm wide. The filaments are pale-yellow (sometimes whitish), dry, and dense with clavate apex. The bauds are globose with slender, dense, stellate greyish hairs. The style apex is flattened-clavate, about 8–9 mm long. The mature capsule is ovate-globose, about 7 mm long and 5.5 mm in diameter, partly glabrous, the style is persistent. The seeds are ovate-turbinate, intensively warty and gibbous, about 1–1.2 mm long and

0.6–0.7 mm wide [11]. Flowering lasts from late May to the end of June, and fruiting from mid-July to early September.

2.2. Species Distribution

Based on the available information, a survey for species appearance was carried out in 2016 and 2017 in a wide area around the three locations recorded in the GBIF (2018) in northeastern Greece. The surveys were carried out on foot by two people in the locations mentioned in the GBIF and in nearby similar areas on the basis of site-related similarities. Special emphasis was given to locations with topographical and ecological characteristics similar to those of the suggested locations for the species, such as altitude, slope aspect, topography, geological and edaphic characteristics, vegetation type and land uses. It should be noted that the reported coordinates of the first species record (in Chrysoupolis) correspond to a flat, agricultural land, 20 m above sea level. Probably, some correction of the coordinates is possible taking into consideration the year of the report (1936).

2.3. Estimation of Environmental Drivers Limiting the Species Distribution

In the locations where the species was found, we recorded the geographical coordination by GPS, the altitude, the topographical characteristics, the slope aspect and inclination, and the soil depth. For local climate estimation, the meteorological data of the nearest meteorological station of Kavala were used, which is at a distance of 5 km from the western area of species distribution and 25 km from the eastern limits, at similar altitude, latitude, and distance from the sea. To estimate the specific soil conditions of the species habitat, a soil sampling was carried out in 2017. Four surface soil samplings were made from four locations, with three replicate samples per location. The soil properties were measured (texture, pH—using the 1:5 (weight/volume) method—total nitrogen, phosphorus, and potassium concentrations) by standard methods for soil analysis [15,16]. In addition, because of the extensive rock presence on the soil surface, a visual estimation of rock percentage covering the sampling area was carried out.

2.4. Community Characteristics and Estimation of Possible Biotic Interactions

To gain a good understanding of the vegetation existing in the species habitat, we used the sampling method of Braun–Blanquet and a specific, modified abundance/dominant scale [17]. Thus, a full record of phytosociological data was made in five plots, sized 100 m^2 [18], in the summers of 2016 and 2018. Floristic elements where collected in the field, while plant taxa identification was made at species level in the laboratory. The plant species found were analyzed according to their functional attitudes, life form, and any possible interactions with *V. dingleri*.

2.5. Fruit and Seed Collection and Laboratory Analysis

We measured the percentage of individuals bearing fruits among 100 randomly selected individual plants and measured the number of fruits for 30 individuals. In early July 2016, only the minimum number of most of the mature fruits was collected (approximately 60–70 fruits from the most productive plants), since in the case of seeds and fruits belonging to rare and protected species, their collection and use in experiments should be limited to minimum [19]. The collected fruits were put in sealed plastic bags, transported to the laboratory (Aristotle University of Thessaloniki) on the same day of collection, and put in a refrigerator. The size (diameter) of all collected fruits was measured using a digital caliper with accuracy of 0.1 mm. Then, they were classified in three size classes, according to their diameter: a large class, with diameter (d) over 3.5 mm, a medium class, with 3.5 mm > d > 3.0 mm, and a small class, with d < 3.0 mm. Afterwards, the seeds were carefully extracted from the fruits and separated from the peel, and the amount of seeds per fruit was measured. The seeds of each fruit were then set separately in small paper bags.

The morphological characteristics of seeds (length, weight, and water content) were determined in a sampling of 15 seeds of five randomly selected fruits per fruit class (225 seeds in total). The seed number per capsule was counted using a stereomicroscope (magnification range 6.7–45×). The floating method was used for seed purity estimation; only high-quality mature seeds were selected for the test. Then, the length and the fresh weight of fully developed seeds were measured in each fruit class. The seed water content was determined following standard laboratory procedures. The seeds were gravimetrically dried at 72 ± 2 °C for 72 h [20], then the final seed water content was calculated on a dry mass basis (%). All seeds were stored in the refrigerator at 4 °C up to the initiation of the germination examinations (three months later).

2.6. Seed Germination under Controlled and Ambient Environmental Conditions

Before assessments, the seeds were surface-sterilized using 0.85% sodium hypochlorite for 1 min, after which they were washed with distilled water. Four replicates of 25 seeds for each of the three fruit classes were placed in glass Petri dishes (9 cm diameter) containing a layer of filter Whatman paper wetted with distilled water. Parafilm M® was used for wrapping the Petri dishes to restrict any moisture loss, while distilled water was added as needed to provide seeds with an adequate moisture level. The Petri dishes were placed in a plant growth chamber at a constant temperature of 20 °C. This temperature was selected on the basis of the existing data for other species of the genus *Verbascum* [21,22]. The fruit size effect on seed germinability was studied by testing the seeds of four fruits per fruit class (1st, 2nd, and 3rd). Seed germination was checked every two days; water was added as needed during the period of the germination test. The criterion for establishing germination was the emergence of a radicle with length of approximately 2 mm [23]. The experiment was terminated when no seeds germinated for one week. The cumulative germination percentage was evaluated every two days, and the final germination after 28 days. The germination percentage was calculated as the average of four replicates of 25 seeds according to Equation (1), and the mean germination time (MGT) was calculated according to Equation (2) [24,25].

$$GP\ (\%) = (\text{number of germinated seeds/total seeds per sample}) \times 100 \quad (1)$$

$$MGT = \Sigma(t.n)/\Sigma n \quad (2)$$

where t is the time (days) from the beginning of the test to the end of the assessment, and n is the number of germinated seeds on day t.

2.7. Seed Germination and Seedling Emergence at Nursery Conditions

Nine fruits were randomly selected (three from each size class), and a random sample of 15 seeds was taken from each of them (in total, 135 seeds). The seeds were planted in plastic pots (Quick pots of 24 cavities with cell volume of 330 cm^3 and depth of 16 cm) in an open-air nursery (research forest nursery of the Laboratory of Silviculture of Aristotle University of Thessaloniki), under relatively similar climatic conditions (similar type of climate, same latitude, closed to the sea), in March 2017. The pots were filled with a common growing medium consisting of peat/perlite in a ratio of 3:1 v/v. The position of the pots in the nursery was changed periodically. All pots were watered to field capacity. After one month, the number of fully developed seedlings (shoot with leaves) per fruit was recorded.

2.8. Statistical Analysis

Statistical analysis of the data was performed using the SPSS software (version 23.0, SPSS Inc., Chicago, IL, USA). Before the analysis, the percentage values of seed germination and seedling emergence were arcsine-transformed to cover the normality and homogeneity assumptions. Seed morphological data as well as the transformed values of seed germination and seedling emergence were subjected to one-way analysis of variance to detect any differences between fruit classes. Comparison of the means followed the least significant differences (LSD) criterion (0.05 level of probability).

3. Results

3.1. Species Geographical Distribution

All locations where the species was observed lie along south-facing, very rocky slopes, at the cliff foot of the mountains Symvolo and Rhodope, at low altitudes (100–150 m asl), approximately 3–15 km from the Aegean Sea, just over the plain (agricultural) area lying between the sea and the mountains. The area of the species appearance is restricted to this altitudinal zone, suggesting that these specific ecological conditions favor its thriving. However, the core population of the species was found more westward with respect to the location indicated in the previous records, at 40°58′16.59″ N, 24°27′54.93″ E, near the village Chalkero, close to the city of Kavala, approximately 150 km from Thessaloniki (Figure 1), in areas exposed to several anthropogenic actions, such as grazing production, livestock raising, agricultural crops cultivation, presence of vehicles, and generally, in degraded habitats.

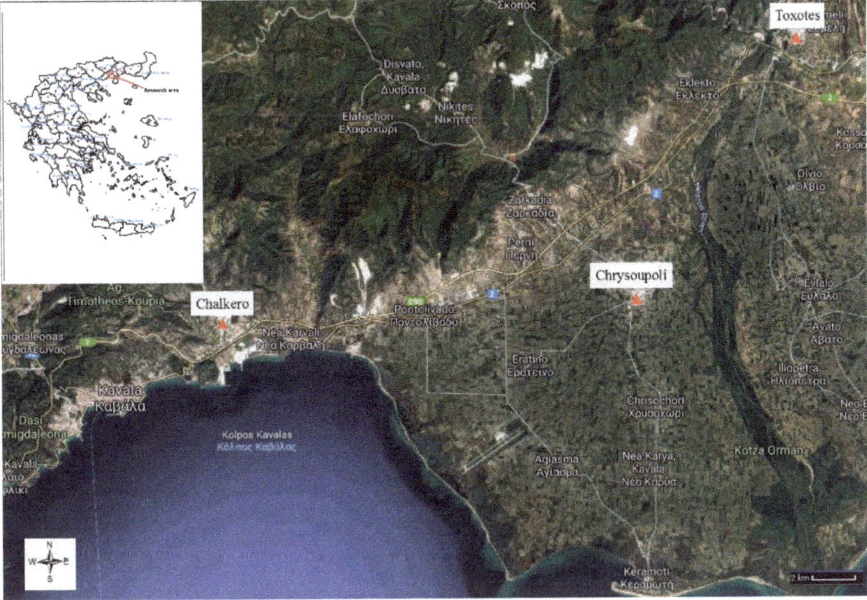

Figure 1. Map showing the location of the species *Verbascum dingleri* occurrence. Geographical coordinates: 40°58′16.59″ N, 24°27′54.93″ E.

The climate of this area belongs to the type Csa according to the Koeppen classification system. On the basis of the available data of the nearest meteorological station of Kavala city, the mean annual temperature is 14.6 °C, and the mean annual rainfall is 602 mm. The mean monthly temperature during the coldest month (January) is 4.2 °C, and that during the warmest month (August) is 26.0 °C, while the prevailing wind is from the southeast (SE) [26].

From a geological point of view, the study area is part of the Rhodope massif that consists of metamorphic and plutonic-eruptive rocks. The specific rock types where the species appears are marbles and limestone. The soils are very shallow (depth range 5–19 cm), with an extremely high rock presence (80–85% of the total area) (Figure 2a, Table 1). The soil is alkaline, and the pH value was found to be almost constant (7.8) in all four sampling points, with very slight differences. However, even though the soil is very shallow, it was found to be relatively rich in organic matter (7.18%), with adequate total nitrogen content (0.42%), as well as phosphorous and potassium contents (8.55 ppm and 336.5 ppm, respectively). Other similar habitats near the species area were investigated during the species flowering period, but we did not locate any additional populations.

Figure 2. Photo of the species *V. dingleri:* (**a**) in situ photo near Chalkero village, showing the dominant site characteristics; (**b**) fruits of *V. dingleri;* (**c**) seeds of *V. dingleri* (photos from a stereomicroscope).

Table 1. Soil characteristics of the habitat of *V. dingleri*.

Soil Sample	Depth Cm	Rock Appearance (%)	Soil Texture	pH	Organic Matter (%)	C (%)	Total N (%)	C/N	P ppm	K ppm
1	13	85	SL	7.76	6.28	3.64	0.315	11.6	4.08	165
2	18	85	SL	7.83	6.86	3.98	0.356	11.2	9.09	320
3	14	80	SL	7.78	5.93	3.44	0.450	7.70	13.75	602
4	19	85	L	7.84	9.63	5.58	0.559	10.0	7.29	259
Mean ± std error	16 ± 1.47	83.8 ± 1.25	SL	7.80 ± 0.02	7.18 ± 0.84	4.16 ± 0.49	0.42 ± 0.054	10.1 ± 0.88	8.55 ± 2.02	336.5 ± 94.07

The site topography provides a habitat that suffers from hot and dry weather conditions during the dry summer season. As summer approaches, the sunrays fall almost vertically, causing the soil to dry and exposing the vegetation to high sunlight. This, in turn, results in high summer evaporation rates, causing the drying of *V. dingleri* and other annual plants and eventually their disappearance during the late summer season.

The spatial data analysis showed that the species niche is determined by the afore-mentioned specific site characteristics that favor the species survival and thriving. The species distribution is probably constrained by hard dispersal obstacles or physiological thresholds along environmental gradients rather than by interactions with other species. Thus, the species appears only in dry, rocky, south-faced slopes, with medium inclination, on shallow soils and limestone, under the climate type Csa, and at a short distance from the sea, in the location at 40°58′16.59″ N, 24°27′54.93″ E.

3.2. Community Characteristics

The vegetation of the area belongs to the Ostryo-Carpinion floristic zone. However, on the basis of the plot where the plant data analysis was made, *V. dingleri* occurs in specific degraded shrub communities, dominated by the woody species *Paliurus spina-christi*, which is a tree species with limited diffusion. These communities consist of a loose shrub layer, dominated by the species *P. spina-christi*, *Olea europea* ssp. *europaea*, and *Quercus coccifera*, with a canopy cover of approximately 30% and a herb layer consisting of the species *Euphorbia dendroides*, *V. dingleri*, *Plantago bellardii*, and many species common in open, degraded, grazing areas of the Mediterranean zone, such as *Avena barbata*, *Allium sphaerocephalon*, *Capsella bursa-pastoris*, and *Bromus tectorum* (the full record of plant abundance/dominance [15] is shown in Table 2). All the recorded species are native, and no exotic species were found. On the basis of the above-mentioned vegetation data, the habitat type where *V. dingleri* was found could be classified as that of the eastern Balkan association *Euphorbio–Paliuretum* or *Oleo sylvestris–Paliuretum spinae christi* [27]. Further data are required for a full phytosociological analysis.

Table 2. Full list of flora taxa recorded in the communities where the species *V. dingleri* was found. The modified abundance/dominant scale according to the Braun Blanquet method [17] was used.

Species	Family	Dominance				
		Plot 1	Plot 2	Plot 3	Plot 4	Plot 5
Euphorbia dendroides	Euphorbiaceae	2a	2m	2m	2b	2a
Paliurus spina-christi	Rhamnaceae	2a	3	2a	2a	2b
Helianthemum sp.	Cistaceae	+		+	r	+
Avena barbata	Poaceae	1	1	+	2m	1
Crupina vulgaris	Asteraceae	R	+			+
Allium sphaerocephalon	Amaryllidaceae	1	1	+	+	1
Planta gobellardii	Plantaginaceae	2m	+	1	1	2m
V. dingleri	Scrophulariaceae	1	+	1	1	1
Asparagus acutifolius	Asparagaceae	2m	1	+	+	1
Melica ciliata	Poaceae	1	+	1	1	1
Capsella bursa-pastoris	Brassicaceae	+	1	+	+	+
Oleaeuropea ssp. europaea	Oleaceae	2b	2a	2b	2b	2a
Quercus coccifera	Fagaceae	2b	2a	2b	2a	2b
Dracunculus vulgaris	Araceae	R				r
Hordeum sp.	Poaceae	1	+		1	+
Bromustectorum	Poaceae	+	+	+	+	+
Vulpia sp.	Poaceae	1	2m		+	1

3.3. Fruit and Seed Diversity

The fruit of *V. dingleri* is a capsule with a diameter of 3–4 mm (mean value 3.24 ± 0.06 mm). The fruits ripen mature from late July to August and then turn green to brown-yellowish. The capsule is generally globose to sub-globose in shape (Figure 2b), and usually two-loculed, with each locule containing approximately half of the fruit's numerous seeds. The average number of fully developed seeds per fruit is 58.18 ± 3.37 (n = 64 fruits), ranging from 15 to 114 (Table 3). Few seeds (3.04 ± 0.86 per fruit) are abnormal (or not fully developed). Statistical analysis revealed significant differences between the three fruit classes in the number of normal seeds and abnormal (or not fully developed) seeds per fruit. In addition, a correlation procedure showed that the number of normal seeds per fruit was significantly positively correlated with the fruit size, i.e., the larger fruit, the higher the number of normal seeds and the lower the number of abnormal seeds (Tables 3 and 4).

Table 3. Fruit and seed characteristics of *V. dingleri* per fruit class. The values are means ± standard error of the mean. Fruit class: Class 1, fruits with diameter (d) > 3.5 mm; Class 2, fruits with 3.5 mm > d > 3.0 mm; Class 3, fruits with d < 3 mm. Means of the same columns followed by different letters are significantly different; ns, non-significant differences.

	Fruit Traits			Seed Traits			
Fruit Class	**Mean Fruit Diameter/mm**	**Number of Mature Seeds**	**Number of Immature Seeds**	**Seed Length/Mm**	**Seed Fresh Weight/gr**	**Seed Dry Weight/gr**	**Seed Moisture Content (%) of Dry Mass**
1st	3.66 ± 0.02a	79.05 ± 4.50a	0.42 ± 0.23c	0.644 ± 0.018 ns	5.0×10^{-4}	3.9×10^{-4}	22.9
2nd	3.26 ± 0.03b	57.04 ± 3.51b	3.92 ± 1.48b	0.626 ± 0.020 ns	4.8×10^{-4}	3.6×10^{-4}	27.8
3rd	2.61 ± 0.08c	29.77 ± 3.88c	5.61 ± 1.43a	0.627 ± 0.033 ns	4.6×10^{-4}	3.4×10^{-4}	29.4
Mean	3.24 ± 0.06	58.18 ± 3.37	3.04 ± 0.86	0.633 ± 0.014	4.8×10^{-4}	3.6×10^{-4}	26.7

Table 4. Results of the statistical analysis for fruit and seeds of *V. dingleri*.

Correlations						
		Fruit Diameter	**Fruit Class**	**N of Normal Seeds**	**N of Abnormal Seeds**	**Germination Final %**
Fruit diameter	Pearson Correlation	1.0	−0.928 **	0.544	0.0 [b]	−0.364
	Sig. (2-tailed)		0.000	0.130		0.335
	N		64	64	64	12
Fruit class	Pearson Correlation			−0.532	0.0 [b]	0.105
	Sig. (2-tailed)			0.141	.	0.787
	N			64	64	12
N of normal seeds	Pearson Correlation				0.0 [b]	0.051
	Sig. (2-tailed)				.	0.896
	N				64	12

Note: ** Correlation is significant at the 0.01 level (two-tailed); [b] cannot be computed because at least one of the variables is constant.

The seeds are very small (minute), ovoid to polygonal, with surface ripples. Mature seeds are dark brown in color (Figure 2c), and their average length is 0.633 ± 0.014 mm. Their mean fresh weight is 4.8×10^{-4} g, which defines them as small seeds. No significant differences were found in seed size between the three fruit classes. The average moisture content of fresh seeds was 26.7% of dry mass, and there was a tendency toward slightly higher values in the larger fruit classes. The seed dispersion of the species is barochory and ornithochory in nature; thus, considering the light seed mass, it is anticipated that species dispersion occurs to some distance from the mother plants.

3.4. Seed Germination Behavior

The germination percentage of *V. dingleri* seeds varied across different fruit classed in laboratory conditions. The best final germination percentage (40%) was observed for smaller seeds belonging to Class 3 (fruit diameter < 3 mm), (Figure 3 and Table 5). Seeds from Classes 1 and 2 showed a quite similar germination pattern, and their final germination percentage was 32% and 26.7%, respectively. However, the fruit size, more specifically, the fruit diameter, does not affect the seed germination percentage of the species. The statistical analysis did not reveal any significant differences among germination percentages. The seeds from all fruit classes started to germinate seven days after sowing, and the germination progress stopped after six weeks. However, most seeds germinated within a period of four weeks, a common period for many species. The mean time to complete germination (MGT) ranged from 17.8 to 24.7 days and was not significantly affected by the fruit class.

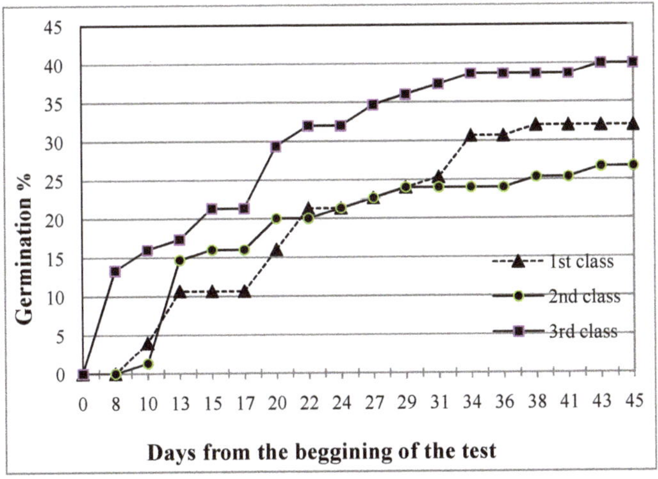

Figure 3. Cumulative germination percentage curves of *V. dingleri* seeds as a function of time for the three fruit classes. Number of seeds: four replications of 25 seeds per fruit class.

Table 5. Seed germination behavior per fruit class in laboratory conditions (final germination percent and mean time to complete germination, MGT) and at nursery (ambient) conditions (percentage of fully developed seedlings) of *V. dingleri*.

	Seed Germination		
	In Laboratory Conditions		At the Nursery (Ambient Conditions)
Fruit Class	**Final Germination (%)**	**Mean Germination Time (MGT)**	**Fully Developed Seedlings (%)**
1st	32.0	24.7	22.0
2nd	26.7	18.8	14.5
3rd	40.0	17.8	9.5
Significance	*ns:* $p > 0.05$	*ns:* $p > 0.05$	
Mean	32.9		15.3

Approximately 30 days after planting, in the middle of spring (April 2017), in ambient conditions, *V. dingleri* germinated seeds produced fully developed seedlings; however, their growth was slow and very similar for all produced plants (statistical analysis did not reveal significant differences among fruit classes, $p > 0.05$) (Table 5). The percentage of fully developed seedlings of *V. dingleri* ranged from 2% to 30% among the fruits, regardless of the fruit diameter.

4. Discussion

4.1. Distribution and Habitat of V. dingleri

The results of this study indicate that the geographical distribution of the species *V. dingleri* lies in the Mediterranean area, at a low altitude of 100–200 m asl, on, generally, south-faced, very rocky slopes of highly degraded, overgrazed areas, in low hill sites. The climate of its habitat belongs to the Csa type, with a mean annual precipitation of 602 mm and a mean annual temperature of 14.6 °C. The species habitat is characterized by very shallow, alkaline soils, rich in organic matter, nitrogen, phosphorus, and potassium. The area consists of a grazing area, close to villages and century-old agricultural land. This means that the species habitat is not an isolated area, but the species co-exists with traditional anthropogenic activities.

From a floristic point of view, the species appears in a specific degraded shrub community where the dominant floristic elements, such as the species *P. spina-christi*, *O. europea ssp. europaea*, and *Q. coccifera*, demonstrate the presence of grazing for a long time. A quite rich herb stratum also appears, consisting of the species *Eu. dendroides*, *V. dingleri*, *P. bellardii*, and many species common in open grazing areas, such as *A. barbata*, *A. sphaerocephalon*, *C. bursa-pastoris*, and *B. tectorum*, which indicates species preference for open grazing, degraded slope areas. According to the inventory data, the species appears locally in a very limited area. Few individuals were recorded in the two years of this study, less than the limit of 500 that is currently considered the minimum number (effective population size) of individuals necessary to secure a species population genetic variability [28,29]. This creates strong uncertainty for the species future perspectives. Such low-sized populations are extremely vulnerable to extinction; however, the available data show that the species has persisted in the area for almost a century (since 1926) and grows between the cities of Kavala and Xanthi, northern Greece, just above the plate agricultural land at the foothills of the mountains, on low-altitude, rocky slopes and very shallow soils. Any specific explanation for this very restricted distribution of the species is still unknown. Probably, this may depend on species regeneration ecology and specific requirements for its propagation in the field. This species' restricted occurrence in a single well-defined area within a small part of the Mediterranean region is a characteristic element of the Mediterranean endemism [30]. Compared with *Verbascum pseudonobile*, another range-restricted *Verbascum* species, which, according to the GBIF, has been recorded in 14 localities in northern Greece, around 41.1° N, 23.624.8° E, it seems that *V. dingleri* is specialized to grow in drier habitats with mild winters, close to the sea (Aegean Sea), while *V. pseudonobile* thrives in colder habitats with a more continental climate. Thus, the species distributions are not overlapping.

4.2. Species Regeneration Ecology

The species fruit contained a quite high number (58.2 ± 3.37) of minute normal seeds. According to the germination experiments, no indication of seed dormancy was observed. Seeds quite highly germinated (up to 80.0%), depending on the fruit, at 20 °C, like many other species of the genus *Verbascum* [21,22]. In addition, the seed germination rate of *V. dingleri* in the open-air nursery, with climatic conditions similar to the natural ones, depended on the fruit from which the seeds were extracted. The seed of some fruits showed a germination of 30%, while the seeds of other fruits did not germinate. The seeds germinated and produced seedlings after approximately 30 days from planting in the middle of spring (April 2017), while their growth was slow and very similar for all produced plants.

4.3. Species Conservation

Both laboratory and nursery results analysis revealed that the seeds of *V. dingleri* do not present dormancy and germinate under favorable environmental conditions (for the species). The seeds of the species showed an average percentage of 30–40% germination at 20 °C, a temperature common in the species habitat area, especially during spring or autumn, when many species regenerate in the fields in temperate zones and with climate type Csa. Thus, we conclude that the species regeneration in the field is theoretically feasible at any time after dispersal. However, in natural conditions, a combination of ecological factors such as temperature, light, and water availability play a key role in regulating plant species germination and seedling emergence. Seeds behavior in response to these crucial environmental factors greatly differs among species, depending on each species eco-physiological attributes [7,8]. Many studies have proven that the temperature greatly influences seed germination, and almost in all species, high temperatures slow down the germination [3,4,31]. Direct sunlight also plays an important role in seed germination behavior [32] as well as in early-stage survival of many non-drought-resistant plant species. Probably, during summer, a combination of high temperatures and high intensity of direct sunlight may be a crucial factor affecting a species regeneration success in the field.

On the other hand, the low amount of precipitation, not only during summer (30–40 mm in the area of the species occurrence, according to the data from the nearest meteorological station of Kavala city), but also during the period of seed dispersion in nature, may be a crucial determining factor for species regeneration. Soil water stress commonly reduces seed germination, since this physiological process is sensitive to water availability, especially during the first stage of germination (swelling of seeds through water adsorption). However, seeds of *V. dingleri* show a tolerance to desiccation, being able to germinate in the presence of a moisture content of approximately 10% (data not shown). Considering that the seeds can survive desiccation, we conclude that *V. dingleri* seeds are orthodox, like the seeds of other species of the genus *Verbascum* [21,22,33]. Thus, in field conditions, it is expected that the tolerance of *V. dingleri* seeds to desiccation could contribute to keep them viable until the time of autumn rains. The findings of the current studies on the seed germination behavior of this species seem not to be able to explain the restricted species occurrence and its narrow endemism. Perhaps, it could be assumed that *V. dingleri* effective regeneration and habitat expansion is determined mainly by seedling propagation ability rather than seed germination in the field. It is worth pointing out that, for the soils of the general area of Kavala city [34,35], some extreme values are recorded for As, Pb, and Zn (which are enriched 7.6, 3.3, and 2.7 times, respectively, in comparison to the values of normal USA soils), even though the majority of the elements in the soils have concentrations within normal ranges. Furthermore, these elements are found at their highest concentrations in the vicinity of the industrial zone of Kavala.

The knowledge of the favorable conditions for early plant growth (germination and seedling emergence) of a rare plant species is definitely necessary for taking the appropriate measures for species conservation and the establishment of population restoration programs [10,36,37]. In the case of endangered species whose habitats are subjected to intensive anthropogenic disturbances, in situ and ex situ species conservation strategies can be suggested, and information about the reintroduction of the species by seedlings produced from seeds collected and effectively treated for germination can be provided [38]. Our study demonstrates that in situ and ex situ conservation and reintroduction of *V. dingleri* using seedlings produced from the seeds collected from a natural population is theoretically feasible.

Specifically, our findings demonstrate that ex situ propagation of *V. dingleri* is feasible from seeds, resulting in the conservation of the species diversity. Thus, these findings can contribute to the advancement of artificial seedlings' production that can be used either for ex situ species conservation or for species reintroduction when the natural population is seriously endangered. However, further research is needed to determine the key factors leading to a satisfactory in situ seedling establishment of this species. More knowledge for effective seed germination and production of high-quality seedlings is crucial for the conservation of this extremely threatened species [9].

Author Contributions: Conceptualization, P.G.; methodology, P.G., M.T., C.D. and A.S.; formal analysis, M.T.; investigation, T.K. (Theodoros Kalapothareas), T.K. (Theodoros Karydopoulos), C.D., A.S. and K.P.; writing—original draft preparation, P.G.; and writing—review and editing, M.T.

Conflicts of Interest: The authors declare no conflict of interest.

References

1. Primack, P.B. *Essentials of Conservation Biology*, 4th ed.; Sinauer Associates: Sunderland, MA, USA, 2006; p. 580.
2. Jusaitis, M.; Polomka, L.; Sorensen, B. Habitat specificity, seed germination and experimental translocation of the endangered herb *Brachycome muelleri* (Asteraceae). *Biol. Conserv.* **2004**, *116*, 251–266. [CrossRef]
3. Shen, S.K.; Wang, Y.H.; Wang, B.Y.; Ma, H.Y.; Shen, G.Z.; Ham, Z.W. Distribution, stand characteristics and habitat of a critically endangered plant *Euryodendron excelsum* HT Chang (Theaceae): Implications for conservation. *Plant Species Biol.* **2009**, *24*, 133–138. [CrossRef]
4. Shen, S.K.; Wang, Y.H.; Ma, H.Y. Seed germination requirements and responses to desiccation and storage of *Apterosperma oblata* (Theaceae), an endangered tree from south-eastern China: Implications for restoration. *Plant Species Biol.* **2010**, *25*, 158–163. [CrossRef]
5. Navarro, L.; Guitián, J. Seed germination and seedling survival of two threatened endemic species of the northwest Iberian Peninsula. *Biol. Conserv.* **2003**, *109*, 313–320. [CrossRef]
6. Cerabolini, B.; De Andreis, R.; Ceriani, R.M.; Pierce, S.; Raimondi, B. Seed germination and conservation of endangered species from the Italian Alps: *Physoplexis comosa* and *Primula glaucescens*. *Biol.Conserv.* **2004**, *117*, 351–356. [CrossRef]
7. Baskin, C.C.; Baskin, J.M. *Seeds: Ecology, Biogeography, and Evolution of Dormancy and Germination*; Academic Press: San Diego, CA, USA, 1998.
8. Fenner, M.; Thompson, K. *The Ecology of Seeds*; Cambridge University Press: Cambridge, UK, 2005; 260p.
9. Butola, J.S.; Badola, H.K. Effect of pre-sowing treatment on seed germination and seedling vigour in *Angelica glauca*, a threatened medicinal herb. *Curr. Sci.* **2004**, *87*, 796–799.
10. Herranz, J.M.; Ferrandis, P.; Martínez-Duro, E. Seed germination ecology of the threatened endemic Iberian *Delphinium fissumsub* sp. *sordidum* (*Ranunculaceae*). *Plant Ecol.* **2010**, *211*, 89–106. [CrossRef]
11. Mattfeld, J. Eine neue Koenigskerze aus Thracien, *Verbascum dingleri*. *Bull. Soc. Bot. Bulg.* **1926**, *1*, 101–103.
12. Global Biodiversity Information Facility (GBIF). Available online: https://www.gbif.org/ (accessed on 31 March 2019).
13. Dimopoulos, P.; Raus, T.; Bergmeier, E.; Constantinidis, T.; Iatrou, G.; Kokkini, S.; Strid, A.; Tsanoudakis, D. *Vascular Plants of Greece: An Annotated Checklist*; Englera; Botanic Garden and Botanical Museum: Berlin, Germany; Hellenic Botanical Society: Athens, Greece, 2013.
14. Flora of Greece Web. 2019. Available online: http://portal.cybertaxonomy.org/flora-greece/intro (accessed on 31 March 2019).
15. Pansu, M.; Gautheyrou, J. *Handbook of Soil Analysis. Mineralogical, Organic and Inorganic Methods*; Springer: Berlin/Heidelberg, Germany, 2007; p. 993.
16. Weil, R.R.; Brady, N.C. *The Nature and Properties of Soils*; Pearson Education Limitted: Harlow, UK, 2017; p. 1083.
17. Barkman, J.J.; Doing, H.; Segal, S. Kritische Bemerkungen und Vorschläge zur quantitativen Vegetationsanalyse. *Acta Bot. Neerl.* **1964**, *13*, 394–419. [CrossRef]
18. Chytrý, M.; Otýpková, Z. Plot sizes used for phytosociological sampling of European vegetation. *J. Veg. Sci.* **2003**, *14*, 563–570. [CrossRef]
19. Graniszewska, M.; Muranyi, R.; Prokopiv, A. Methods of germination and cryogenic storage of rare species seeds from the Ukrainian Carpathians. *Visnyk L'viv Univ. Ser. Biol.* **2004**, *36*, 153–159.
20. ISTA (International Seed Testing Association). International rules for seed testing. *Seed Sci. Technol.* **1999**, *13*, 229–513.
21. Senel, E.; Ozdener, Y.; Incedere, D. Effect of temperature, light, seed weight and GA3 on the germination of *Verbascum bithynicum*, *Verbascum wiedemannianum* and *Salvia dicroantha*. *Pak. J. Biol. Sci.* **2007**, *10*, 1118–1121. [PubMed]
22. Seipel, T.; Alexander, J.M.; Daehler, C.C.; Rew, L.J.; Edwards, P.J.; Dar, P.A.; McDougall, K.; Naylor, B.; Parks, C.; Pollnac, F.W.; et al. Performance of the herb *Verbascum thapsus* along environmental gradients in its native and non-native ranges. *J. Biogeogr.* **2015**, *42*, 132–143. [CrossRef]
23. Ganatsas, P.; Tsakaldimi, M.; Zachariadis, G. Effect of air traffic pollution on seed quality characteristics of *Pinus brutia*. *Environ. Exp. Bot.* **2011**, *74*, 157–161. [CrossRef]

24. Ganatsas, P.; Tsakaldimi, M.; Thanos, C. Seed and cone diversity and seed germination of *Pinus pinea* in Strofylia Site of the Natura 2000 Network. *Biodivers. Conserv.* **2008**, *17*, 2427–2439. [CrossRef]
25. Martínez-Díaz, E.; Martínez-Sánchez, J.J.; Conesa, E.; Franco, J.A.; Vicente, M.J. Germination and morpho-phenological traits of *Allium melananthum*, a rare species from south-eastern Spain. *Flora* **2018**, *249*, 16–23. [CrossRef]
26. Hellenic National Meteorological Service. *Climatic Data of the Greek Network, Period 1930–1975*; Hellenic National Meteorological Service: Athens, Greece, 1978. (In Greek)
27. Casavecchia, S.; Biscotti, N.; Pesaresi, S.; Edoardo Biondi, E. The *Paliurus spina-christi* dominated vegetation in Europe, 2015. *Biologia* **2015**, *70*, 879–892. [CrossRef]
28. USDA Forest Service. *An Interim Guide to the Conservation and Management of Pacific Yew*; USDA Forest Service: Portland, OR, USA, 1992.
29. Piovesan, G.; Saba, E.P.; Biondi, F.; Alessandrini, A.; Di Filippo, A.; Schirone, B. Population ecology of yew (*Taxus baccata* L.) in the Central Apennines: Spatial patterns and their relevance for conservation strategies. *Plant Ecol.* **2009**, *205*, 23–46. [CrossRef]
30. Thompson, J.D. *Plant Evolution in the Mediterranean*; Oxford University Press: Oxford, UK, 2005; 302p.
31. Núñez, M.R.; Calvo, L. Effect of high temperatures on seed germination of *Pinus sylvestris* and *Pinus halepensis*. *For. Ecol. Manag.* **2000**, *131*, 183–190. [CrossRef]
32. Schütz, W.; Rave, G. The effect of cold stratification and light on the seed germination of temperate sedges (*Carex*) from various habitats and implications for regenerative strategies. *Plant Ecol.* **1999**, *144*, 215–230. [CrossRef]
33. Catara, S.; Cristaudo, A.; Gualtieri, A.; Galesi, R.; Impelluso, C.; Onofri, A. Threshold temperatures for seed germination in nine species of *Verbascum* (Scrophulariaceae). *Seed Sci. Res.* **2016**, *26*, 30–46. [CrossRef]
34. Papastergios, G.; Filippidis, A.; Christofides, G.; Kassoli-Fournaraki, A.; Fernández–Turiel, J.L.; Georgakopoulos, A.; Gimeno, D. Trace elements in uncultivated surface soils in the Kavala area, northeastern Greece. In Proceedings of the 11th International Congress, Athens, Greece, 24–26 May 2007.
35. Papastergios, G. Environmental Geochemical Study of Soils and Sediments in Coastal Areas, East of Kavala (Macedonia, Greece) and Production of Geochemical Maps via the Use of GIS. Ph.D. Thesis, Aristotle University of Thessaloniki, Thessaloniki, Greece, 2008; 224p. (In Greek, with English summary).
36. Walck, J.L.; Baskin, J.M.; Baskin, C.C. Ecology of the endangered species *Solidago shortii*. VI. Effects of habitat type, leaf litter, and soil type on seed germination. *J. Torrey Bot. Soc.* **1999**, *126*, 117–123. [CrossRef]
37. Cirak, C. Seed germination protocols for ex situ conservation of some *Hypericum* species from Turkey. *Am. J. Plant Phys.* **2007**, *2*, 287–294.
38. Shen, S.; Wang, Y.; Zhang, A.; Fuqin, W.; Jiang, L.; Jiang, L. Conservation and reintroduction of a critically endangered plant *Euryodendron excelsum*. *Oryx* **2013**, *47*, 13–18. [CrossRef]

Article

Characterization of Small Forest Landowners as a Basis for Sustainable Forestry Management in the Libertador General Bernardo O'Higgins Region, Chile

Francisca Ruiz-Gozalvo [1], Susana Martín-Fernández [2,*] and Roberto Garfias-Salinas [1]

1 Department of Forest Management and Environment, University of Chile, Santiago 11315, Chile; franciscaruiz86@gmail.com (F.R.-G.); rgarfias@uchile.cl (R.G.-S.)
2 Forestry School, Technical University of Madrid, 28040 Madrid, Spain
* Correspondence: susana.martin@upm.es; Tel.: +34-91-336-6401

Received: 22 October 2019; Accepted: 12 December 2019; Published: 16 December 2019

Abstract: Sclerophyllous forests are extremely sensitive to global warming, and the sclerophyllous forest in the possession of small forest landowners (SFLs) in the Libertador General Bernardo O'Higgins Region in Chile is degraded in spite of their high ecological value. Due to the total lack of forest management, the yield obtained from native forests is very low, with highly intervened forests and intense soil erosion. The main contribution of this article is to present, for the first time, a study on the characterization and problems of 211 small forest landowners in this region of Chile. After interviewing the landowners, multivariate analysis techniques were applied to the results of the survey, which enabled four types of SFL to be identified. Differences were found in regard to the surface area of their properties and the products extracted, among others. However, they all had a similar social profile, low education level and little training in forest management, very advanced ages, a lack of initiative to create forest communities, and lack of basic services due to their isolation. The characterization of the SFLs allowed proposals to be designed for future sustainable forest management activities to help mitigate the continuous deterioration of the native forest and obtain products in a sustainable way and with greater yields, considering current legal aspects, access to subsidies, and specific forest training plans for each type of SFL.

Keywords: small forest landowner; sclerophyllous forest; sustainable forest management; multivariate analysis

---

1. Introduction

Sustainable forest management (SFM) is the most widely known type of management at the global level. SFM is supported by policies and legislation in 97% of forest areas worldwide, and the highest values are found in South and East Asia and Central and South America, with between 93% and 100% [1]. Two of the main tools used to incentivize SFMs are stakeholder involvement and forest management plans. However, the percentage of forest areas with forest management plans varies by region, ranging from below 20% in South America and West and Central Africa to over 90% in Europe and Central America [1].

In this context, as a country participating in the workgroup on criteria and indicators for the Conservation and Sustainable Management of Temperate and Boreal Forests (Montreal Process), Chile has carried out a line of work aimed at developing support tools for the monitoring and assessment of the sustainability of forest management in subnational projects, conducting studies in areas with humid climates and mild summers in the regions of Araucania, Los Lagos, and Biobio (see Figure 1).

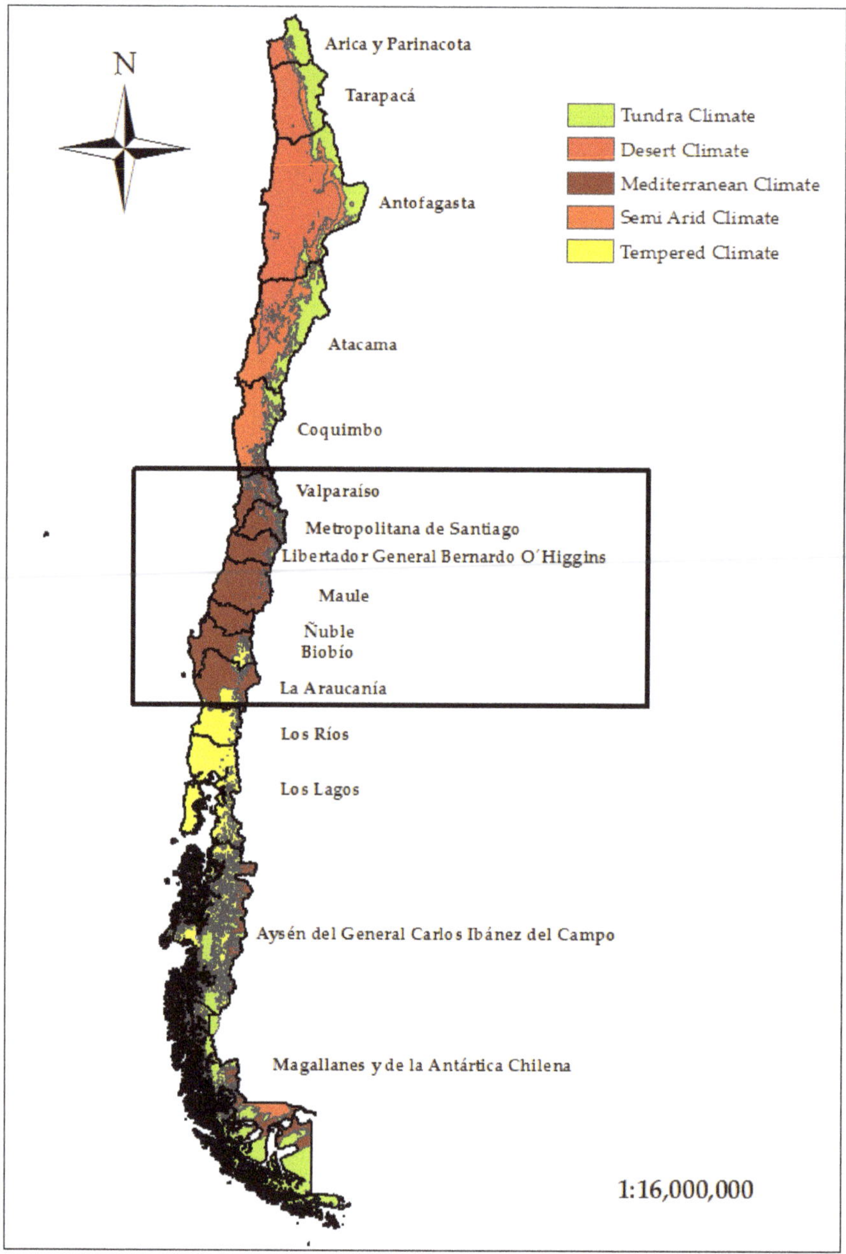

Figure 1. Climate types in the regions in Chile; the Mediterranean climate area is framed.

However, this line of work has not been applied to sclerophyllous forests, which today are far from their climax [2] in spite of being located in Chile's central region—the most developed and densely populated in the country—on the slopes of the Coastal Range and Los Andes and representing one of the world's diversity hotspots [3]. These forests are the most vulnerable to climate change, which increases their risk of land degradation and desertification [4]. According to the latest climate model simulations at the global level, global warming could increase dryland ecosystems from 11% to 23% of Earth's land surface by the end of the 21st century [4,5]. Although Chile has improved its subsidy requirements to favor SFLs, the lack of forest management plans in this region means that SFLs do not receive subsidies that could encourage landowners [6].

To protect the remaining native vegetation and possibly reverse its declining trend, it is essential to implement programs for sustainable management with the participation of the state and the private sector.

The VI Libertador General Bernardo O'Higgins Region is part of the central zone of Chile and is characterized by having extensive areas of this forest type, and by the fact that a large proportion of the forests are in the hands of small forest landowners [7]. The total area of native forests is 485,790 ha—49,742 ha belong to 2957 SFLs, 413,126 ha to other private landowners, and 22,922 ha to protected areas, according to the 2013 Agroforestry Census developed by Instituto de Desarrollo Agropecuario de Chile. The application of the forest legislation may change these figures, but there is still no updated census.

The native forests in the possession of small forest landowners (SFL) are associated to other productive systems, such as agriculture and livestock farming, generating a wide variety of production units in the region. One characteristic of these subsystems is their low yields, mainly used for self-consumption, while any surplus goes for sale or exchange in order to ensure the subsistence and food security of the family group [8,9].

Forest areas are mainly managed by their owners using their empirical skills and knowledge of the resource. However, these interventions have not been sufficient to maintain the productivity of the native forest [2]. In addition, activities such as changes in land use for agricultural purposes, the utilization of the native forest as a source of refuge and food for cattle, and the constant extraction of forest products without any sustainable management have contributed to reducing the tree cover and generated processes of forest fragmentation [8,9].

Elsewhere, the successful implementation of sustainable actions requires both the local communities and the decision-makers to understand the ecological, environmental, and cultural dynamics and the productive potential of the forests [10].

Small forest landowners are a group with diverse social, economic, and productive characteristics. The lack of information on this population sector hinders the processes of decision making and the orientation of promotional instruments and support programs [8]. Their correct characterization can contribute to the design of forestry, agricultural, and livestock policies that facilitate the transfer of technology [11] and the development, implementation, and monitoring of rural development projects [12].

Classification is an important tool for reducing heterogeneity and complexity in planning. According to Carmona and Nahuelhual (2009) [13], "A typology is a way of conceptualizing this reality and allows the resources to be directed more effectively and efficiently. As a result, units of relation can be formed between decision-makers and their natural environment. It facilitates the spatial-temporal observation of the effects of their decisions at the level of the landscape and territory, and the results can be used to supplement future interventions according to the characteristics of the actors existing in it."

Previous work on the classification of rural family units and their associated property systems can be found in the literature. However, this classification has been done based on objective information and assessments, so the studies tend to be descriptive rather than explanatory or predictive [14,15].

Multivariate analysis methods offer objective classification techniques such as principal components analysis, multiple correspondence analysis and cluster analysis, which can group together the landowners with homogeneous characteristics in order to enable decisions to be made in a relatively similar way to allow the visualization, analysis, and understanding of the current productive systems so they can be given similar recommendations [12,13,16]. These techniques can be complemented with local expert knowledge [16] to identify sustainable lines of work as a basis for improving the focus of programs that promote scale economies and regional development [17].

The aim of this study is to characterize and typify small forest landowners in the Libertador General Bernardo O'Higgins Region using objective information and the participation of the local population, in addition to multivariate statistical techniques to identify problems and deficiencies and propose future lines of work that lead to the sustainable management of the territory and to rural development.

2. Materials and Methods

2.1. Study Area

The study area (see Figure 2) is located in the VI Libertador General Bernardo O'Higgins Region, Chile, between 33°51′02″ and 34°56′36″ S, and between 72°00′12″ and 69°48′38″ W. It has an area of 16,387 km^2, and a population of 914,555 in 2017 [18]. The average altitude is 251 m a. s. l. from sea level to an altitude of 4500 m. The local relief is divided into four characteristic sectors—the Los Andes Range, the intermediate depression, the Coastal Range, and the coastal plain. It has a predominantly Mediterranean climate characterized by rainy winters and dry summers. Average annual precipitation is 680 mm, with variations caused by the local topography. The climate on the coast is cloudy with abundant humidity, whereas in the interior, there are significant temperature variations of over 13 °C [19].

Figure 2. Map of the study area.

The plant diversity is high in terms of its composition, structure and conservation status. There is a predominance of tall shrubs with sclerophyllous leaves, along with low-growing xerophytic and spiny shrubs, succulents and very tall sclerophyllous, spiny laurifoliate trees. The dominant species include litre (*Lithraea caustica (Molina)* Hook et Arn), quillay (*Quillaja saponaria Molina*), peumo (*Cryptocarya alba (Molina) Looser*), boldo (*Peumus boldus Molina*), and espino (*Acacia caven Molina*) [19].

2.2. Data Collection

This analysis was begun by consulting the Native Forest Registry, which provided information on the forest type and the dominant species present in the study area. It should be noted that the map scale contemplated in the databases was in many cases insufficient, so this information was supplemented with validation campaigns on the terrain in the region that currently has native forest.

From the information available in the registry, an analysis was made of the communities in 2015, including the commune and property limits, toponymy, paths and population centers. The properties that complied with the restrictions imposed by Law 20.283 on the Recovery of the Native Forest and Forest Development in Chile were identified in each community. This document defines the characteristics of the SFLs as follows:

> "A person who has the title deed for one or more rural properties with the presence of native forest, with a combined area of no more than 200 ha; whose assets do not exceed the equivalent of 3500 development units (accounting unit used in Chile to re-value savings in line with inflation rates so the money maintains its purchasing power); equivalent to 122,500 euros, whose income derives mainly from agricultural or forestry activities and who works directly on the land, on their property or on another property belonging to third parties."

This information was supplied by the following territorial institutions: National Forestry Corporation (CONAF), a local development program (PRODESAL), and municipalities in the region, who together directed the subsequent campaigns on the terrain for the survey of socio-economic information by first visualizing the spatial distribution of the small forest landowners in the region.

This information was used to prepare a census and a basic map to incorporate and analyze the properties and check whether they meet the legal requirement. There is currently no census of small forest landowners in Chile, as prescribed by this law.

The properties that met the requirements of this study were localized through campaigns on the terrain, involving visits to all the small landowners and their properties. Productive, social, and economic information was collected on the small forest landowners in the region. All the interviews were conducted with the head of the household.

Semi-structured interviews were carried out to capture primary information, on the form included in Appendix A. This procedure was chosen due to the open character of this kind of survey and the fact that they allow information to be obtained from the interviewee in a fluid way [20].

Finally, the information was incorporated into a database for its subsequent statistical analysis.

2.3. Statistical Analysis

The semi-structured interview provided information on 41 variables, from which were selected the most representative variables in the study. The data processing was done with IBM SPSS Statistics software version 20 (Armonk, NY, USA).

The variables were chosen according to the value of the coefficients of variation (CV) for each initial variable, selecting the variables whose CV was greater than or equal to 50% of variability in terms of the mean [21].

The multivariant method of multiple correspondence analysis (MCA) was applied to the selected variables. This method analyzes the relation between categories of quantitative variables so their dimension can be reduced. Finally, Ward's method of hierarchical cluster analysis was applied to define homogeneous groups of SFLs in the study area according to their similarities for the variables ultimately selected [22].

2.4. Validation of Typologies and Work Lines

This study was conducted within the framework of the project "Program for training and technological transfer for a better application of Law 20.283, aimed at small landowners in the VI Region." In this project, the problems, needs, and training of the local communities were discussed with local leaders, and the SFL typologies were obtained. However, forest activities were proposed to each landowner according to the characteristics of the property and in compliance with Law 20.283.

In addition, each typology was validated and the landowners' problems and future lines of work were identified by consulting a panel of experts described in Table 1.

Table 1. Experts consulted for the validation of typologies and lines of work proposed.

Specialty	Institution
Forestry Engineer; Diploma of Specialization in Silviculture	University of Chile
Forestry Engineer; Doctor in Forestry Science	University of Chile
Veterinarian; Doctorate in Agrarian Economy	University of Chile
Veterinarian; Technical head of PRODESAL; Specialist in rural development	Local development program, Lolol
Agricultural technician; Technician PRODESAL; Specialist in rural development	Local development program, Chépica
Forestry Engineer; Head of Forestry Department	National Forestry Corporation
Forestry Engineer; Forest Extension Specialist	National Forestry Corporation

3. Results and Discussion

3.1. Spatial Distribution of Small Forest Landowners in the Region

The analysis of the maps, supplemented with information from territorial institutions, initially identified 420 small forest landowners in the region, which represented 14% of the total SFLs, according to the 2013 Agroforestry Census. The 420 small forest landowners were natural persons and not companies or associations. In the interview it was verified that the property met the requirements established by Law 20.283 in its definition of the characteristics of the small forest landowner. This verification eliminated any landowner who did not fulfill any of the following requirements: the property did not have native forest, the area was larger than 200 ha, the landowners did not have the deed to the property, or the information collected in the survey was incomplete. Finally, the number of SFLs was reduced to 211. Average area of native forest in this sample was 18.25 ha, and the sampling relative error was 16%.

A useful map was generated for the subsequent classification of the producers based on the number of properties that meet the initial criteria for selection and analysis for each commune.

Table 2 shows the community to which the SFLs belong, the associated area of native forest and the number of landowners. Figure 3 shows their location. As can be seen in Figure 3, no landowners can be seen in the eastern and western zone (Coastal Range and Los Andes Range) of the map as the forests in these zones are not sclerophyllous.

Table 2. Spatial distribution of small forest landowners and area of native forest at the commune level.

Commune	Number of SFLs	Area of Native Forest (ha)	Commune	Number of SFLs	Area of Native Forest (ha)
Chépica	37	545.3	Nancagua	5	207
Coinco	1	8	Navidad	6	41.5
Coltauco	7	114.8	Palmilla	5	34.5
Doñihue	12	260.3	Peralillo	12	119.3
La Estrella	9	238	Pichidegua	11	203
Las Cabras	2	61	Placilla	8	211.3
Litueche	14	135.3	Pumanque	15	403.25
Lolol	28	702	Quinta de Tilcoco	1	10
Machalí	4	48.1	Requínoa	1	0.5
Malloa	6	61.8	San Vicente	8	79.5
Marchigüe	11	260	Santa Cruz	9	293.36
Total SFLs		211	Total area		4035.81

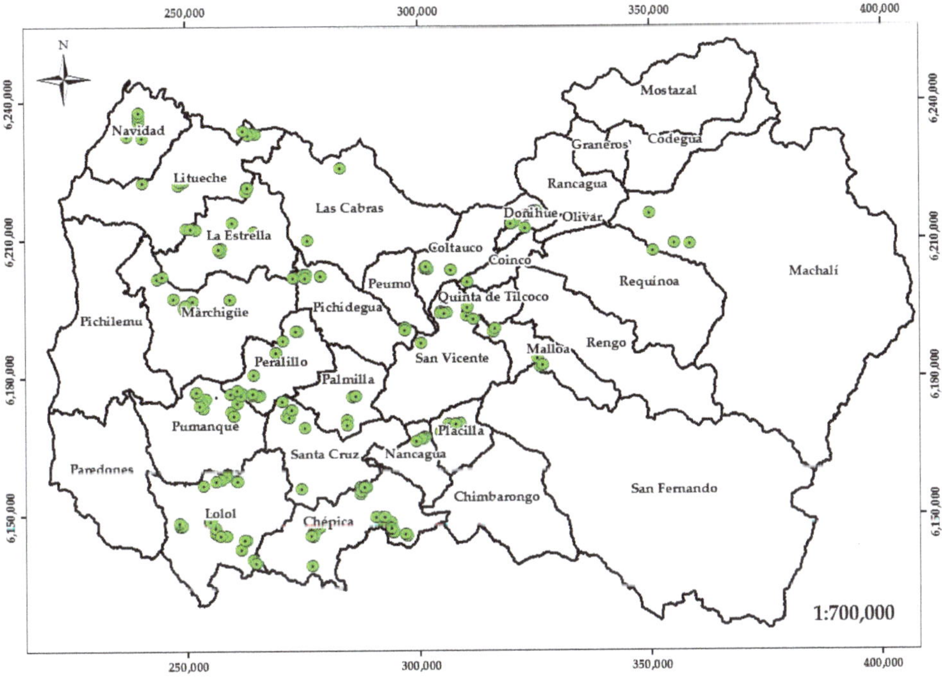

Figure 3. Spatial distribution of small forest landowners (SFL) in the Libertador General Bernardo O'Higgins Region.

3.2. *Identification of Landowner Groups*

3.2.1. Initial Variables

The choice of variables was made according to their discriminant power, based on their coefficient of variation and prior studies. Of the total of 14 variables selected, four correspond to quantitative variables and ten to categorical variables. The selected quantitative variables were age of the head of household (AG), property area (PA), area of native forest (NFA), and per capita income (PCI). The selected qualitative variables were type of animal unit (AU), economic activity of the head of household (EA), forest status (FS), education of the head of household (ED), infrastructure (IN), extraction of forest products (FP), main problem (MP), main productive subsystem (MPS), training subject (TR), and source of water for agriculture and livestock (WS).

3.2.2. Results of the Multiple Correspondence Analysis

The multiple correspondence analysis separated three dimensions that explain 60% of the total variance, each with similar inertia values of between 18.4% and 21.3%.

Table 3 shows the discriminatory capacity of each variable in each dimension. As can be seen, Dimension 1 is mainly explained by productive variables (main productive subsystem, type of animal unit, and source of water for agriculture and livestock); Dimension 2 is explained by property variables (property area and area of native forest); and Dimension 3 is explained by social variables (age, main economic activity, and education of the head of household).

Table 3. Discriminant measures for each variable.

Variable	Dimension			Mean	Variable	Dimension			Mean
	1	2	3			1	2	3	
AG	0.257	0.145	0.416	0.273	AU	0.375	0.249	0.049	0.224
EA	0.320	0.230	0.363	0.304	FS	0.106	0.024	0.056	0.062
ED	0.187	0.204	0.260	0.217	IN	0.181	0.190	0.086	0.152
PA	0.211	0.431	0.240	0.294	FP	0.062	0.132	0.010	0.068
NFA	0.153	0.364	0.267	0.261	MP	0.145	0.085	0.145	0.125
MPS	0.477	0.273	0.155	0.302	TR	0.208	0.187	0.244	0.213
WS	0.268	0.183	0.185	0.212	PCI	0.320	0.230	0.363	0.304

3.2.3. Grouping of Landowners

The cluster analysis, through Ward's method, identified small landowners with similar characteristics in the three dimensions (eight variables) obtained from the MCA—property, productive, and social. Based on this, four homogeneous groups of producers were formed: elderly SFLs, retired SFLs, largest SFLs, and middle-aged SFLs, as shown in Table 4.

Table 4. Description of typologies based on variables obtained from the multiple correspondence analysis (MCA).

Variables	Typologies			
	Elderly SFLs	Retired SFLs	Largest SFLs	Middle-Aged SFLs
Number of SFLs	93	34	15	69
1. Average age of the head of household	71	60	65	52
2. Education level	Basic level, not completed	Basic level, not completed	Basic level, not completed	Basic level, completed
3. Average property area (ha)	28	11	117	15
4. Average area of native forest (ha)	15	8	95	11
5. Main economic activity	Full-time work on the property	Retired	Full-time work on the property	Full-time work on the property
6. Main productive subsystem	Livestock	None	Livestock	Livestock
7. Type of animal unit	Sheep	None	Sheep	Sheep
8. Water for agriculture and livestock	Well	None	Well	Well

The reduction of the sample size from 420 SFL to 211 implied a smaller sample size for each typology and a greater sampling error. The highest standard deviation (SD) for the different variables and typologies in Table 6 corresponded to largest SFLs, with 15%, while the maximum SD for the other typologies was 5% for the elderly SFLs, 8.5% for retired SFLs, and 9.2% for middle-aged SFLs. In further studies, it will be necessary to increase the sample size of the SFL with the largest properties.

The characterization of producers was supplemented with information on specific variables related to forest management, agricultural production, and characteristics of the family groups obtained in the survey in order to refine the profile of each type of producer and observe some of the typical problems in each group (see Tables 5 and 6).

Table 5. Characteristics of the family group.

Typology	Family Members and Activities
Elderly SFLs	52%, 1 or 2 members: activities related to the property. 48%, over 2 members: 23% students; 13% engaged in activities related to the property; 64% other activities
Retired SFLs	56%, 1 or 2 members: retired. 44%, over 2 members: 23% students; 6% engaged in activities related to the property; 23% other full-time activities
Largest SFLs	66%, 1 or 2 members: activities related to the property 34%, over 2 members: 100% engaged in activities related to the property;
Middle-aged SFLs	39%, 1 or 2 members: activities related to the property. 61%, over 2 members: 25% students, 16% engaged in activities related to the property.

Table 6. General observations for each typology.

Variables	Typology			
	Elderly SFLs	Retired SFLs	Largest SFLs	Middle-Aged SFLs
SFLs who extract forest products: %/no./SD	63/59/0.05	44/22/0.085	73/11/0.11	50/35/0.06
Main species	*Acacia caven*	*Acacia caven* and *Peumus boldus*	*Acacia caven* and *Peumus boldus*	*Acacia caven*
Main product extracted	Firewood	Firewood	Firewood, charcoal, boldo leaves	Firewood
Destination of the production	Self consumption	Self consumption	Self consumption and commercialization	Self consumption
Forestry management plan	No	No	No	No
Forest structure	Mature with scarce regeneration	Mature with scarce regeneration	Mature with scarce regeneration	Mature with scarce regeneration
Forest status	Highly intervened Intense soil erosion	Minimally intervened, low soil erosion	Highly intervened in some sectors	Highly intervened in some sectors
Forest activity performed by SFLs	Clearing	Clearing	Nothing	Clearing
Agricultural production	Private orchard	Private orchard	Private orchard	Private orchard
Equipment	Animal traction	Nothing	Mechanically powered	Animal traction
Destination of the production	Self consumption	Self consumption	Self consumption	Self consumption
Minimum distance to paved road	10 km	10 km	10 km	10 km
Problems	Availability of water	Availability of water Connectivity	Availability of water Basic services	Availability of water
SFLs with training: %/no./SD	66/61/0.05	34/12/0.08	60/9/0.15	44/30/0.06

Table 6. *Cont.*

Variables	Typology			
	Elderly SFLs	Retired SFLs	Largest SFLs	Middle-Aged SFLs
SFLs trained in the subject of forestry: %/no./SD	6/6/0.024	18/6/0.07	30/5/0.12	15/10/0.09
SFLs participating in social organizations: %/no./SD	83/77/0.04	61/21/0.08	73/11/0.12	73/50/0.05
Social organization	Neighborhood association	Neighborhood association	Neighborhood association	Neighborhood association
Properties entered in the property register: %/no./SD	71/66/0.047	56/19/0.08	67/10/0.14	62/43/0.06
Employees	0	0	0	0
Per capita income	$8–$150	$8–$150	$238–$426	$8–$150

By typifying the small forest landowners in the region, it was possible to recognize their characteristics and the problems that need to be tackled in a specific way for each situation. This diagnostic is essential for designing programs or lines of work. The detailed analysis of the variables revealed the similarities and differences between the four typologies of small forest landowners in the Libertador General Bernardo O'Higgins Region.

Similar behavior was observed in the characteristics of the head of household, namely advanced average age and low education level. Most landowners in the typologies elderly SFLs, retired SFLs, and largest SFLs are illiterate. Middle-aged SFLs, the youngest group (average 52 years), have completed at least the basic education level, and 63% of the university graduates interviewed belong to this group.

The main economic activity among elderly SFLs and largest SFLs is full-time work on the property, and the livestock subsystem accounts for the largest proportion of household income. This is followed by the agricultural subsystem, whose yield is used for the consumption of the family group. In the case of the retired SFLs, the producers' income comes mainly from their retirement, and although the main economic activity in the middle-aged SFLs typology is full-time work on the property (41%), 39% of the producers carry out activities outside the property on either a full-time or seasonal basis. They are also quite isolated. The closest paved road is at least 10 km away, which hinders their access to markets and services.

The size of the property is the main difference between typologies from the economic point of view. SFLs in the largest SFLs typology have the largest properties (117 ha) on average, with an average of 93 ha of native forest. The largest SFLs typology has the lowest number of SFLs, with 15 out of a total of 221, indicating the dispersion of properties in the native forest, which is a disadvantage for forest management.

This is the only group that rents mechanical traction equipment; it also has the highest per capita income, at between $238 and $426, and the highest number of owners who extract forest products, including firewood, charcoal, and boldo leaves. These are the only producers who allocate part of the products extracted from the forest for commercialization, and 50% have received training in the field of forestry, although the total number of trained people is close to the other typologies. This typology has the highest number of family members involved in the management of the property, which ensures their future engagement with forest management and production. However, there are no data on the volume of the products extracted for any of the typologies. Firewood is used as a source of energy for the home and for cooking, which explains the interest among the SFLs in the restoration of native forests, as this would represent a source of savings [9,23,24].

Regarding the status and management of the native forest, the landowners have very little training in forest management, with only 27 SFLs trained in the subject of forestry. The largest SFLs typology, with the highest average size, has the highest proportion and the best forest training. The only forest activity the landowners perform is clearing. According to the answers in the survey, they consider this activity to be the correct way to manage the forest. This lack of knowledge of forest management, coupled with the fact that none of the native forest properties had a forestry plan, leads to a substantial rate of intervention in the forest, which causes little regeneration and soil erosion problems. In addition, since property passes down through families, many new landowners are not entered in the property registry. This is one of the main problems when applying for subsidies and reduces the capacity for production, forest conservation, and water management [8,9]. These problems also have a global impact, as these landowners play an important role in preventing the deforestation of the native forest, which has great potential for carbon mitigation [25].

In this context, there is a unanimous need for all the typologies to implement sustainable management plans whose activities could be subsidized (Law 20.283 on the Recovery of the Native Forest and Forest Development), and for training in forest management, the role of the native forest, and the market for and commercialization of timber and non-timber forest products.

This would represent a first step toward the recovery of these forests, increase water availability, and encourage forest development among landowners [6]. The activities identified according to the landowners' typology and the use of the forest, validated by the panel of experts, are shown in Table 7.

Table 7. Eligible activities proposed by typology.

	Typology			
	Elderly SFLs	Retired SFLs	Largest SFLs	Middle-Aged SFLs
Objective	Timber production	Timber production	Timber production	Timber production
Main species	*Acacia caven*	*Acacia caven* and *Peumus boldus*	*Acacia caven* and *Peumus boldus*	*Acacia caven*
Activities	-Cleaning, pruning, thinning -Supplementary planting	-Regenerative felling -Natural regeneration -Cleaning, pruning, thinning	-Cleaning, pruning, thinning -Supplementary planting -Infiltration ditches -Protection against forest fires	-Cleaning, pruning, thinning -Supplementary planting

The SFLs' only type of organization is the neighborhood association to which they belong. There is no kind of SFL community forest management in the region that promotes the conservation of forests, ensures the landowners' income from the use of the forest, and improves the governance of the management [26,27].

There is therefore a lack of resources to transform these SFLs into the main agents of their development, intervening in the decision-making processes together with other stakeholders (technicians, administration, policy-makers), and avoiding the generation of policies at the territorial level that have a generalizing tendency and whose impact is unclear [23,28]. This would increase the efficiency of their own potential and of the public and private initiatives that affect their development [9,28].

As a line of work for local management, we therefore propose the creation of social innovation and knowledge dissemination networks that make it possible to design, operate and assess strategies to stimulate innovation among the landowners. These networks are systems of informal interrelation that can be easily disassembled and recombined, encouraging non-hierarchical relations of trust between their members (SFLs, other landowners, technicians, Administration), and may endure over time [29]. In the case of the SFLs in the VI Region, after characterizing the landowners and identifying the forestry technology practices and innovations to be adopted, the key local actors can be identified due to their greater knowledge of the forestry activity or higher social prestige. These key local actors catalyze the processes of dissemination and adoption of innovations [30].

4. Conclusions

The main contribution of this article is to present for the first time a study on the characterization and the problems of small forest landowners of native sclerophyllous forests in the Libertador General Bernardo O'Higgins Region.

Sclerophyllous forests are extremely sensitive to global warming, and the sclerophyllous forest in the hands of small forest landowners in this region of Chile is degraded in spite of its high ecological value. Due to the total lack of forest management, the yield obtained from the native forests is very low.

Although the characterization of small forest landowners reveals clear differences in terms of the surface area of their properties, the vast majority have a similar social profile, a low education level, and very advanced ages. This information is important for devising the training strategy to be used. The main product extracted is firewood for self-consumption for the household, leading to various degrees of intervention in the resource.

This classification allowed the design of proposals for activities for sustainable management in the future, promoting regeneration either naturally or through planting, and fundamentally clearing and thinning activities, in order to reduce the continuous degradation of the native forest while considering the current legal aspects and the access to subsidies, all supported by specific forestry training plans for each typology.

It is therefore necessary to develop participative political processes that ensure rural development, the creation of associations of forest landowners, training, and the sustainable management of these forest areas. This should include the measurement of criteria and indicators in order to monitor their ecological, economic and social development, and the incentivization of research into areas of sclerophyllous forest.

Finally, the methodology presented in this study, with its quantitative focus, can be replicated in sectors where there is no information available on small forest landowners, in order to generate areas for recommendation that consider the needs of the local population and ensure they are not merely passive actors in their own development processes.

Author Contributions: Conceptualization, F.R.-G.; R.G.-S., and S.M.-F.; Data collection, F.R.-G. and R.G.-S. Formal analysis, F.R.-G. and S.M.-F.; Writing—Original draft, F.R.-G. and S.M.-F.; Writing—Review and editing, S.M.-F. and R.G.-S.

Funding: This research received no external funding.

Acknowledgments: We would like to thank the project "Program for training and technological transfer for a better application of Law 20.283, aimed at small landowners in the VI Region" financed by the Research Fund into the Native Forest at the National Forest Corporation and executed by the University of Chile.

Conflicts of Interest: The authors declare no conflict of interest.

Appendix A

Table A1. Interview applied to small forest landowners in the Libertador General Bernardo O'Higgins Region.

FORM no.	DATE:		SURVEYOR:	
	COMMUNE		UTM N	
	LOCALITY		UTM E	
	ADDRESS		DISTANCE TO PAVED ROAD	
	ALTITUDE		CONTACT	
	SLOPE			
IDENTIFICATION OF THE HEAD OF HOUSEHOLD				
NAME	*AGE*	*MARITAL STATUS*	*MAIN ECONOMIC ACTIVITY*	*EDUCATION*

Table A1. *Cont.*

CHARACTERISTICS OF THE FAMILY GROUP				
NAME	*AGE*	*MARITAL STATUS*	*MAIN ECONOMIC ACTIVITY*	*EDUCATION*

CHARACTERISTICS OF THE PROPERTY		CHARACTERIZATION OF THE NATIVE FOREST	
LANDHOLDING		*STRUCTURE*	ADULT FOREST
OWN (ROLE)			RENEWAL
SUCCESSION			ADULT FOREST RENEWAL
			ADULT FOREST SCRUB
LEASED		*MAIN SPECIES*	
RIGHT TO USUFRUCT			SCLEROPHYLLOUS
FREE CONCESSION			HUALO OAK
			CHILEAN CEDAR
SIZE OF THE PROPERTY			CHILEAN PALM
AREA OF NATIVE FOREST		*FOREST TYPE*	EVERGREEN, OTHER

CHARACTERIZATION OF PRODUCTIVE SUBSYSTEMS (Indicate the Main One)			
AGRICULTURAL SUBSYSTEM (Indicate areas)		*LIVESTOCK SUBSYSTEM*	
FARMLAND		*FOREST SUBSYSTEM*	
VINES		*AGROFORESTRY SUBSYSTEM*	
FRUIT TREES		*OTHER*	

WATER AVAILABILITY INDICATE WHETHER THE WATER FOR CONSUMPTION AND IRRIGATION COMES FROM A WELL, WATERWHEEL, ETC.	

Table A1. *Cont.*

YIELDS. EXTRACTION OF TIMBER AND NON-TIMBER FOREST PRODUCTS. INDICATE WHICH ONES AND WHAT QUANTITY	
NUMBER AND TYPE OF ANIMAL UNITS	
ANIMAL-DRAWN AND MECHANICAL TRACTION EQUIPMENT.	
GENERAL INFRASTRUCTURE (DESCRIBE)	

Source: Adapted from [7].

References

1. MacDicken, K.G.; Sola, P.; Hall, J.E.; Sabogal, C.; Tadoum, M.; Wasseige, C. Global progress toward sustainable forest management. *For. Ecol. Manag.* **2015**, *352*, 47–56. [CrossRef]
2. Castillo, M.; Molina, J.R.; Rodríguez y Silva, F.; García-Chevesich, P.; Garfias, R. A system to evaluate fire impacts from simulated fire behavior in Mediterranean areas of Central Chile. *Sci. Total Environ.* **2016**, *579*, 1410–1418. [CrossRef] [PubMed]
3. Armesto, J.J.; Arroyo, M.; Hinojosa, L.F. The Mediterranean environment of Central Chile. In *The Physical Geography of South America*; Velben, T.T., Young, K.R., Orme, A.R., Eds.; Oxford University Press: New York, NY, USA, 2007; pp. 184–199.
4. Bastin, J.F.; Berrahmouni, N.; Grainger, A.; Maniatis, D.; Mollicone, D.; Moore, R.; Patriarca, C.; Picard, N.; Sparrow, B.; Abraham, E.M.; et al. The extent of forest in dryland biomes. *Science* **2017**, *356*, 635–638. [CrossRef]
5. Huang, J.; Yu, H.; Guan, X.; Wang, G.; Guo, R. Accelerated dryland expansion under climate change. *Nat. Clim. Chang.* **2016**, *6*, 166–171. [CrossRef]
6. Cubbage, F.; Harou, P.; Sills, E. Policy instruments to enhance multi-functional forest management. *For. Policy Econ.* **2007**, *9*, 833–851. [CrossRef]
7. Garfias, R.; Soto, M.C.; Gozalvo, F.R.; Alonso, A.V.; Intveen, H.B.; Cerrillo, R.N. Remants of schlerophyllous forest in the mediterranean zone of central Chile: Characterization and distribution of fragments. *Interciencia* **2018**, *43*, 655–663.
8. Bahamondez, A.; Rivas, E. Comunidades vinculadas a los Recursos Vegetacionales. In *Bosque Nativo, Comunidades y Cambio Climático*; Bahamondez, A., Ed.; European Union Commission: Brussels, Belgium, 2016; pp. 87–112.
9. FAO. *State of the Forests in the World: Forest Pathways to Sustainable Development*; Food and Agriculture Organization of United Nations: Roma, Italy, 2018; p. 153.
10. Estrada, C.; Irisity, F.; Hernández, G.; Torres, J. *Casos Ejemplares de Manejo Forestal Sostenible en Chile, Costa Rica, Guatemala y Uruguay*; Food and Agriculture Organization of United Nations: Roma, Italy, 2016; p. 236.
11. Aquino, V.; Camarena, F.; Julca, A.; Jiménez, J. Multivariate characterization of tarwi (Lupinus mutabilis Sweet) producing farms of the Mantaro Valley, Peru. *Sci. Agropecu.* **2018**, *9*, 269–279. [CrossRef]

12. Kuivanen, K.; Álvarez, S.; Michalscheck, M.; Adjei-Nsiah, S.; Descheemaeker, K.; Mellon-Bedi, S.; Groot, J.C. Characterising the diversity of smallholder farming systems and their constraints and opportunities for innovation: A case study from the Northern Region, Ghana. *NJAS Wagening. J. Life Sci.* **2016**, *78*, 153–166. [CrossRef]
13. Carmona, A.; Nahuelhual, L. Tipificación y caracterización de sistemas prediales: Caso de estudio en Ancud, Isla de Chiloé. Classification and characterization of farming systems: Case study in the municipality of Ancud, Chiloé Island. *Agro Sur* **2009**, *37*, 189–199. [CrossRef]
14. González-Redondo, P.; Delgado-Pertíñez, M.; Toribio, S.; Ruiz, F.A.; Mena, Y.; Caravaca, F.P.; Castel, J.M. Characterisation and typification of the red-legged partridge (Alectoris rufa) game farms in Spain. *Span. J. Agric. Res.* **2013**, *8*, 624–633. [CrossRef]
15. Köbrich, C.; Rehman, T.; Khan, M. Typification of farming systems for constructing representative farm models: Two illustrations of the application of multi-variate analyses in Chile and Pakistan. *Agric. Syst.* **2003**, *76*, 141–157. [CrossRef]
16. Mądry, W.; Gozdowski, D.; Roszkowska-Mądra, B.; Castel, J.M.; Mena, Y.; Hryniewski, R. An overview of farming system typology methodologies and its use in the study of pasture-based farming system: A review. *Span. J. Agric. Res.* **2013**, *11*, 316–326. [CrossRef]
17. Carrillo Anzures, F.; Acosta Mireles, M.; Flores Ayala, E.; Torres Rojo, J.M.; Sangerman-Jarquín, D.M.; González, L.; Buendía, E. Caracterización de productores forestales en 12 estados de la República Mexicana. *Rev. Mex. Cienc. Agric.* **2017**, *8*, 1561–1573. [CrossRef]
18. Censo de Chile. 2017. Available online: http://www.censo2017.cl/ (accessed on 21 September 2019).
19. Región del Libertador Bernardo O'Higgins. Available online: https://www.bcn.cl/siit/nuestropais/region6/ (accessed on 20 June 2019).
20. Vega, N. La Entrevista Como Fuente De Información: Orientaciones Para Su Utilización. In *Memoria e Historia del Pasado Reciente. Problemas Didácticos y Disciplinares*; Alonso, L., Falchini, A., Eds.; Secretaría de Extensión, Universidad Nacional del Litoral: Santa Fe, Argentina, 2009; pp. 51–77.
21. Wei, X.; Allen, N.J.; Liu, Y. Disparity in organizational research: How should we measure it? *Behav. Res. Methods* **2016**, *48*, 72–90. [CrossRef] [PubMed]
22. Martín-Fernández, S.; Martinez-Falero, E. Sustainability assessment in forest management based on individual preferences. *J. Environ. Manag.* **2018**, *206*, 482–489. [CrossRef] [PubMed]
23. Herrmann, T.M.; Torri, M.C. Changing forest conservation and management paradigms: Traditional ecological knowledge systems and sustainable forestry: Perspectives from Chile and India. *Int. J. Sustain. Dev. World Ecol.* **2009**, *16*, 392–403. [CrossRef]
24. Li, Y.; Mei, B.; Linhares-Juvenal, T. The economic contribution of the world's forest sector. *For. Policy Econ.* **2019**, *100*, 236–253. [CrossRef]
25. Watson, R.T.; Noble, I.R.; Bolin, B.; Ravindranath, N.H.; Verardo, D.J.; Dokken, D.J. *Land Use, Land-Use Change, and Forestry, Intergovernmental Panel on Climate Change (IPCC) 2000*; Cambridge University Press. New York, NY, USA, 2000; p. 388.
26. Agrawal, A.; Chhatre, A.; Hardin, R. Changing governance of the world's forests. *Science* **2008**, *320*, 1460–1462. [CrossRef]
27. Newton, P.; Schaap, B.; Fournier, M.; Cornwall, M.; Rosenbach, D.W.; DeBoer, J.; Whittemore, J.; Stock, R.; Yoders, M.; Brodnigc, G.; et al. Community forest management and REDD+. *For. Policy Econ.* **2015**, *56*, 27–37. [CrossRef]
28. Lee-Cortés, J.; Delgadillo-Macías, J. Territorial potential as factor of development, rural management model. *Agric. Soc. Desarro.* **2018**, *15*, 191–213.
29. Aguilar, J.; Altamirano, J.; Rendón, R. *Del Extensionismo Agrícola a las Redes de Innovación Rural 2010*; Universidad Autónoma de Chapingo: Texcoco, Mexico, 2010; p. 282.
30. Aguilar Gallegos, N.; Olvera Martínez, J.A.; Martínez González, E.G.; Aguilar Ávila, J.; Muñoz Rodríguez, M.; Santoyo Cortés, H. La intervención en red para catalizar la innovación agrícola. *Redes Rev. Hisp. Anal. Redes Soc.* **2017**, *28*, 9–31.

Article

Soil Organic Carbon Accumulation in Post-Agricultural Soils under the Influence Birch Stands

Tomasz Gawęda [1,*], Ewa Błońska [2] and Stanisław Małek [3]

1 State Forests National Forest Holding, Bielsko Forest District, 43-382 Bielsko–Biała, Poland
2 Department of Forest Soil Science, Faculty of Forestry, University of Agriculture, 31-425 Krakow, Poland
3 Department of Forest Ecology and Reclamation, Faculty of Forestry, University of Agriculture, 31-425 Krakow, Poland
* Correspondence: tgtomik@poczta.onet.pl

Received: 7 July 2019; Accepted: 4 August 2019; Published: 8 August 2019

Abstract: The aim of this study was to demonstrate the effects birch renewal on the soil organic carbon accumulation and on dehydrogenase activity. We selected 12 research plots with birch stands of different ages (1–4 years, 5–8 years, 9–12 years, and 13–17 years) to determine soil texture, pH, total carbon and nitrogen levels, and base cation content. The total organic carbon stock was calculated for the soil profiles. Additionally, dehydrogenase activity was determined. Naturally regenerated birch stands on post-agricultural land facilitated carbon accumulation. Based on our results, dehydrogenase activity is useful in assessing the condition of post-agricultural soils, and its determination allowed for us to assess the processes occurring in post-agricultural soils that are associated with the formation and carbon distribution.

Keywords: afforestation; birch stands; carbon stock; dehydrogenase activity

---

1. Introduction

Poland has one of the largest forest areas in Europe; forests occupy 29% of the country's territory and cover an area of 9.1 million hectares [1]. Forest stands on arable lands occupy nearly 25% of the forest area. The soil and soil processes are crucial in maintaining the productivity of forest ecosystems [2]. Soil is an important reservoir of carbon, and it is estimated that the global soil carbon stocks amount to more than 1500 Pg C, which are significantly higher than those of the atmosphere (750 Pg C) or the biomass of terrestrial ecosystems (650 Pg C) [3]. Recently, the mechanisms that are responsible for carbon stabilisation in soils have received considerable interest due to their relevance in our understanding of the global carbon cycle [3]. The fertility and productivity of soils depend on soil organic matter (SOM), which serves as a nutrient reservoir and it therefore plays an important role in nutrient cycling [4]. Changes in soil management are the main factor that affects SOM dynamics [5]. For example, transforming natural ecosystems into arable fields generally depletes soil organic carbon (SOC) reserves by as much as 75% (mostly between 30 and 50%), depending on the climate zone and the ecosystem type [6]. However, such losses can be limited by converting arable land into grassland and forest [7]. Afforestation positively affects various soil properties, and the soil organic carbon values of afforested sites are generally higher than those of bare sites [8]. Afforestation also positively influences the physical soil characteristics, which are important for maintaining soil stability and productivity [9–11].

In recent years, the interest in soil quality has been stimulated by the growing awareness of the fact that the soil is an important component of the biosphere. It functions not only to produce food, timber, and other forest resources, but it also plays a role in maintaining the local, regional, and global quality of the environment. Karlen et al. [12] and Gil-Sotres et al. [13] state that soil quality enables the healthy functioning of an ecosystem and it maintains its biological production. One of the most important factors in determining soil fertility are the soil biological properties in relation to the activity of microorganisms and higher organisms (plants and animals), including enzymes that are secreted by them [14]. Dehydrogenase activity, as an integral part of an intact cell and soil microflora activity, can provide information regarding the biologically active population of microorganisms in a given soil [15]. Soil microbial and enzymatic activity responds relatively quickly to slight changes in soil conditions and can reflect the changes in soil quality before they can be detected by other soil analyses [16]. Dehydrogenase plays a significant role in the biological oxidation of soil organic matter by transferring hydrogen from organic substrates to inorganic acceptors [17]. In this sense, the determination of dehydrogenase activity can be used to reflect the changes in soil biology [18,19], including assessing soil quality, the influence of soil management on soil quality, and the degree of regeneration of degraded soil [13,20]. Afforestation induces a rapid increase in microbial biomass, with changes apparent within one year of tree planting [21]. In a previous study, afforestation increased bacterial PLFAs by 20–120%, whereas it had a stronger impact on the development of fungal communities (increases by 50–200%) [22].

In this context, the main aim of this research was to determine the effects of changes in soil management from agriculture to forestry on the soil organic carbon accumulation and on enzymatic activity. Dehydrogenase activity, which plays a key role in the carbon cycle, was determined, and we tested the following hypotheses: (1) natural birch regeneration has a positive effect on the soil organic carbon accumulation and (2) dehydrogenase activity reflects the changes that occurred in the soil of the studied chronosequence.

2. Materials and Methods

The soil samples were collected from 12 research plots at four locations in the Mazowieckie province of Poland (Table 1, Figure 1). The study area is characterized by the following climatic conditions: average annual rainfall of 629 mm, average annual temperature of 8.4 °C, and a growing season of 210 days. The area in which the sample plots were located was dominated by fluvioglacial and glacial sand and loam with Gleysols, Cambisols, Podzols, and Arenosols [23]. The study plots were used as cropland in the past.

The study plots were divided into four groups based on the age of the self-seeded birch trees: I—1–4 years, II—5–8 years, III—9–12 years, and IV—13–17 years. In each plot, we took three soil samples from the 0–5, 5–15, and 15–50 cm layers. The samples were air-dried, sieved through a 2-mm-mesh, and the following physicochemical properties were determined [24]: pH (potentiometrically, in 1 M KCl and H_2O solution), texture (using laser diffraction in an Analysette 22: Fritsch, Idar-Oberstein, Germany), nitrogen, and organic carbon contents (with a LECO CNS True Mac Analyser: Leco, St. Joseph, MI, USA), C/N ratio, basic cations content (in 1 M ammonium acetate, using a Thermo Scientific iCAP 6000 ICP OES analyser, Thermo Fisher Scientific, Cambridge, UK). The data presented is the mean of the three soil replicates.

The results were used to calculate the carbon stock in the soils of the chronosequences, based on bulk density (BD), which were determined using Kopecky's cylinders. The carbon stock was calculated according to the following formula:

$$\text{SOCstock} = \text{C} \times \text{BD} \times \text{T}/100 \quad (1)$$

where SOCstock is the carbon stock in the soil ($kg{\cdot}m^{-2}$), C is the carbon content in the soil layers ($g{\cdot}kg^{-1}$), BD is bulk density [$g{\cdot}cm^{-3}$], and T is the thickness of the soil layers (cm).

Fresh samples, with natural moisture content, were taken to determine dehydrogenase (DH) activity (DH) via the Lenhard method, according to the Casida procedure. The DH activity was expressed as μmol TPF kg^{-1} h^{-1} [25].

The biomass [kg·ha^{-1}] of the aboveground and belowground parts of the stands in groups I–IV was determined, using the trunks, branches, assimilation apparatus, bark, and roots. For this, 10 trees were randomly selected at each location and were separated into trunk, branches, assimilation apparatus, bark, and roots. All the parts of the tree were weighed in the field while using portable scales with an accuracy of 0.01 g. Samples from each of the components from each tree model were collected to determine the relationship between fresh and dry biomass. Briefly, the samples were oven-dried at 105 °C and then weighed. On the basis of appropriate fresh-to-dry mass ratios, we calculated the dry biomass of the components for each tree.

Basic statistical data were calculated (i.e., the arithmetic mean and measures to determine the degree of differentiation among the results—standard deviation). The obtained data did not show normality, the Shapiro–Wilk test was used to to check the normal distribution. Tukey's HSD multiple comparisons of means were used in post hoc analysis to assess the effect of the age of regenerated birch trees and soil depth on the studied soil properties. Principal components analysis (PCA) was used to interpret the relationships among the studied variables, while the Pearson's correlation was applied to determine the relationships between dehydrogenase activity and soil properties. By applying Ward's method, the samples were grouped according to DH activity and carbon content. Average and standard deviation (SD) were presented in tables. Differences with $p < 0.05$ were considered to be statistically significant. Statistical analyses were performed in the Statistica 10 software (StatSoft Inc., Tulsa, OK, USA).

Table 1. Location of research plots and soil type.

Study Site	GPS	Soil Type
Mińsk Maz.	52°10′ N, 21°40′ E	Brunic arenosol
Kozienice	51°24′ N, 21°26′ E	Brunic arenosol
Dobieszyn 1	51°35′ N, 21°10′ E	Brunic arenosol
Dobieszyn 2	51°33′ N, 21°09′ E	Brunic arenosol

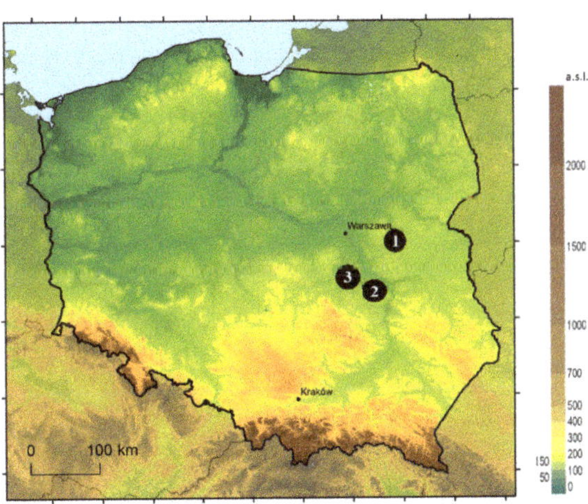

Figure 1. Localization of study plots (1—Mińsk Maz., 2—Kozienice, 3—Dobieszyn 1, and Dobieszyn 2).

3. Results

The soils differed in terms of pH values. The highest average pH was recorded in soils from the youngest birch stands (groups I and II); in the surface soil layer of these stands, the pH in H_2O was 4.52 and 4.62, respectively. The lowest pH was recorded in soils of the oldest stands (group IV) (Table 2).

All of the sites were similar in terms of silt, and clay contents; there were no statistically significant differences of the studied chronosequences. Slight differences were noted in the sand content (Table 2). We also found no statistically significant differences in the C contents of the subsequent soil layers. There were no significant differences in the rate of organic matter decomposition, being expressed as the C/N ratio. The highest C/N ratio was recorded for the soil of the youngest birch stands (group I, average 22.3) and the lowest for the soil of the oldest birch stands (group IV, average 16.6). There were statistically significant differences in the Ca content (Table 2).

The total carbon stock did not significantly differ among the groups (Table 3). A slightly lower than average carbon stock was found in soils of the younger stands (groups I and II), while the values were above the average in the soils of the older stands (groups III and IV). However, these differences were not statistically significant. The average carbon stock depended on the age of the forest stand and it ranged from 4.34 to 6.16 kg·m^{-2}. The carbon stock in the different soil layers changed with the age of the tree stands (Table 3). In the soils of the younger stands (groups I and II), a greater amount of accumulated carbon was found in the upper layer (0–5 cm) as compared to the same layer in the older stands (groups III and IV). In the soils of groups I and II, the proportion of the total carbon, which was determined to a depth of 50 cm, in the surface layer accounted for about 28%, while it accounted for 13.6% in the soils of the oldest stands. The highest amount of C in the deeper layers was recorded in soils of the oldest stands; the carbon in the 15–50-cm soil layer of group IV accounted for nearly 60% of the total carbon stock, while it did not exceed 40% in the soils of group I.

Dehydrogenase activity was used as a proxy for the biological activity of the studied soils and it varied among the sites. The highest mean value of dehydrogenase activity was recorded for group I soils and the lowest for group II–IV soils (Figure 2), which indicated a decrease in dehydrogenase activity with stand age. A strong relationship between dehydrogenase activity and the basic cation content was determined while using Pearson's correlation coefficient (Table 4). The correlation coefficients between dehydrogenase activity and K, Ca, and Mg contents were 0.81, 0.64, and 0.60, respectively.

Table 5 presents the components of the aboveground and belowground biomass of the examined stands. The biomass components significantly increased with stand age. In the youngest stand (group I), the average aboveground biomass was 2521.5 kg·ha^{-1} and the root biomass was 1058.7 kg·ha^{-1}. In the oldest stand (group IV), the aboveground biomass was 30 times higher than that in the youngest stand, whereas the root biomass was more than 12 times higher in the older group IV than in the younger (group I) stand.

The first two axes of the PCA explained 46.2% of the variance of the analyzed soil properties (Figure 3). The first axis explained 31.74% of the variability and it was mainly related to the basic cation content, while the second axis explained 14.47% of the variability and it was associated with the C and N contents and with the pH. The results of the PCA analysis confirmed the dependence of dehydrogenase activity on the amounts of basic cations that are available. The C and N levels were higher in the soils of the older stands. To discriminate the distinction of the studied chronosequence of birch stands, we performed a cluster analysis, which enabled us to identify the two main groups differing in dehydrogenase activity and carbon content. The youngest stands (groups I and II) clearly differed from the oldest stands (groups III and IV) (Figure 4).

Table 2. Soil properties of the studied chronosequence of birch stands, with statistic test results.

Chronosequence	Depth	pH in H_2O	pH in KCl	C	N	C/N	Na	K	Ca	Mg	Sand	Silt	Clay
				%			cmol (+)·kg^{-1}				%		
I	0–5	4.52 ± 0.14^{a}	3.88 ± 0.12^{a}	1.57 ± 0.33^{a}	0.07 ± 0.02^{a}	22.3 ± 6.5^{a}	0.17 ± 0.05^{a}	0.95 ± 0.30^{a}	1.62 ± 0.21^{ab}	1.57 ± 0.57^{a}	83 ± 4^{a}	14 ± 3^{a}	2 ± 1^{a}
	5–15	4.73 ± 0.30^{ab}	3.97 ± 0.19^{a}	1.17 ± 0.71^{a}	0.07 ± 0.03^{a}	14.1 ± 4.9^{a}	0.21 ± 0.07^{a}	1.05 ± 0.47^{a}	1.60 ± 0.57^{ab}	1.60 ± 0.70^{a}	84 ± 4^{ab}	13 ± 3^{a}	2 ± 1^{a}
	15–50	5.03 ± 0.48^{a}	4.23 ± 0.16^{a}	0.49 ± 0.49^{a}	0.03 ± 0.02^{b}	19.0 ± 16.2^{a}	0.25 ± 0.07^{a}	1.50 ± 0.40^{a}	1.69 ± 0.56^{a}	2.60 ± 0.85^{a}	79 ± 11^{a}	18 ± 7^{a}	3 ± 2^{a}
II	0–5	4.62 ± 0.31^{a}	3.98 ± 0.07^{a}	1.30 ± 0.31^{a}	0.09 ± 0.07^{a}	15.2 ± 6.5^{a}	0.22 ± 0.10^{a}	1.98 ± 0.30^{a}	2.13 ± 0.71^{a}	2.23 ± 1.39^{a}	76 ± 2^{b}	19 ± 2^{a}	4 ± 1^{a}
	5–15	4.87 ± 0.23^{a}	4.[illegible]3 ± 0.26 [a]	1.10 ± 0.47^{a}	0.32 ± 0.35^{a}	11.1 ± 2.7^{a}	0.21 ± 0.10^{a}	1.39 ± 1.34^{a}	2.37 ± 1.11^{a}	2.30 ± 1.48^{a}	78 ± 5^{b}	18 ± 4^{a}	4 ± 2^{a}
	15–50	5.29 ± 0.38^{a}	4.35 ± 0.39^{a}	0.16 ± 0.10^{a}	0.04 ± 0.03^{ab}	8.4 ± 5.3^{a}	0.22 ± 0.07^{a}	1.96 ± 0.34^{a}	2.25 ± 1.01^{a}	3.55 ± 0.31^{a}	80 ± 5^{a}	17 ± 4^{a}	3 ± 1^{a}
III	0–5	4.58 ± 0.15^{a}	3.79 ± 0.12^{a}	1.38 ± 0.58^{a}	0.11 ± 0.06^{a}	15.8 ± 7.2^{a}	0.20 ± 0.10^{a}	0.80 ± 0.25^{a}	1.38 ± 0.14^{b}	1.44 ± 0.32^{a}	83 ± 7^{ab}	14 ± 5^{a}	2 ± 2^{a}
	5–15	4.60 ± 0.20^{ab}	3.94 ± 0.15^{a}	1.19 ± 0.63^{a}	0.06 ± 0.04^{a}	19.7 ± 4.9^{a}	0.19 ± 0.06^{a}	0.81 ± 0.27^{a}	1.41 ± 0.18^{a}	1.51 ± 0.37^{a}	86 ± 4^{ab}	12 ± 3^{a}	2 ± 1^{a}
	15–50	4.91 ± 0.36^{a}	4.27 ± 0.05^{a}	0.68 ± 0.53^{a}	0.05 ± 0.05^{ab}	18.2 ± 15.1^{a}	0.21 ± 0.05^{a}	1.07 ± 0.55^{a}	1.35 ± 0.24^{ab}	1.92 ± 0.52^{a}	87 ± 3^{a}	12 ± 2^{a}	2 ± 1^{a}
IV	0–5	4.38 ± 0.13^{a}	3.85 ± 0.25^{a}	1.39 ± 0.50^{a}	0.09 ± 0.06^{a}	16.6 ± 6.2^{a}	0.21 ± 0.03^{a}	1.11 ± 0.18^{a}	1.66 ± 0.31^{ab}	1.97 ± 0.27^{a}	85 ± 7^{ab}	13 ± 5^{a}	2 ± 1^{a}
	5–15	4.35 ± 0.30^{b}	3.93 ± 0.09^{a}	1.33 ± 0.55^{a}	0.15 ± 0.06^{a}	11.5 ± 2.7^{a}	0.23 ± 0.03^{a}	1.10 ± 0.26^{a}	1.41 ± 0.19^{b}	1.81 ± 0.38^{a}	88 ± 1^{a}	11 ± 1^{a}	1 ± 1^{a}
	15–50	4.65 ± 0.59^{a}	4.13 ± 0.29^{a}	1.11 ± 0.90^{a}	0.19 ± 0.02^{a}	12.2 ± 8.9^{a}	0.22 ± 0.08^{a}	1.19 ± 0.33^{a}	1.54 ± 0.09^{a}	2.22 ± 0.58^{a}	88 ± 2^{a}	11 ± 2^{a}	2 ± 1^{a}

Mean ± SD; small letters in the upper index of the mean values mean significant differences of soils properties between chronosequence and dept.

Table 3. Total and percentage carbon storage ($kg \cdot m^{-2}$—SOCstock) in soil layers of the studied chronosequence of birch stands.

Chronosequence	Depth	SOCstock ($kg \cdot m^{-2}$)	Total SOCstock in All Layers	% Participation SOCstock
I	0–5	0.94	4.57	27.5
	5–15	1.43		32.2
	15–50	2.20		40.2
II	0–5	0.78	4.34	24.1
	5–15	1.37		38.4
	15–50	2.19		37.5
III	0–5	0.82	5.41	18.0
	5–15	1.49		29.1
	15–50	3.10		52.8
IV	0–5	0.69	6.16	13.6
	5–15	1.54		27.7
	15–50	3.93		58.6

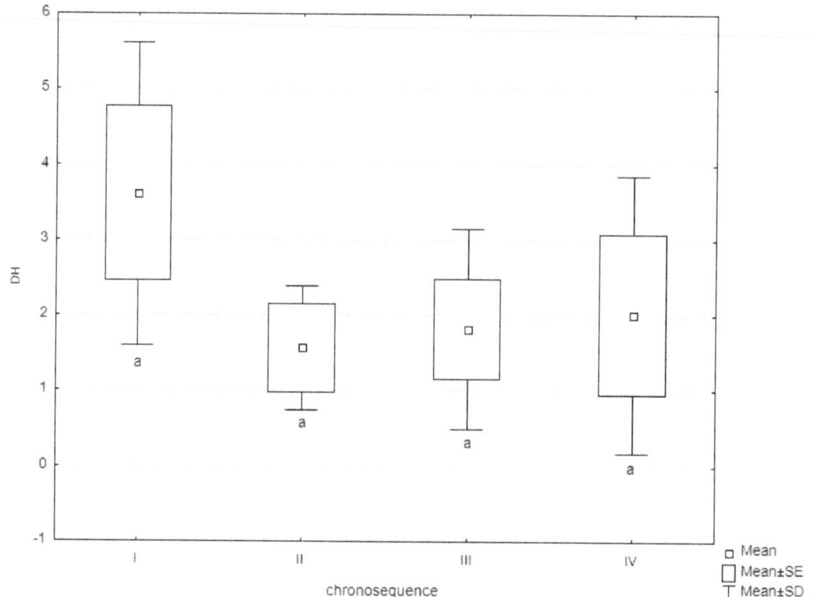

Figure 2. Dehydrogenase activity (DH) ($\mu mol\ TPF \cdot kg^{-1} \cdot h^{-1}$) in first soil layers of the studied chronosequences of birch stands.

Table 4. Correlations between dehydrogenase activity (DH) and basic soil properties.

	pH_{H2O}	pH_{KCl}	Na	K	Ca	Mg	C	N	Sand	Silt	Clay
DH	−0.04	−0.13	0.43	0.81 *	0.64 *	0.60 *	0.09	0.10	−0.20	0.20	0.11

* $p < 0.05$.

Table 5. Average biomass (kg·ha^{-1}) of stand components in the studied chronosequence.

Chronosequence	Stem	Branches	Foliage	Bark	Roots
I	867.6 ^{b}	548.0 ^{b}	863.5 ^{b}	242.4 ^{b}	1058.7 ^{b}
II	7947.7 ^{ab}	1924.9 ^{b}	1745.9 ^{ab}	1931.4 ^{ab}	2866.8 ^{b}
III	23907.2 ^{ab}	5147.2 ^{ab}	2299.7 ^{ab}	4581.3 ^{ab}	6485.4 ^{ab}
IV	54307.9 ^{a}	11357.8 ^{a}	3242.6 ^{a}	9173.7 ^{a}	13492.7 ^{a}

Small letters in the upper index of the mean values mean significant differences.

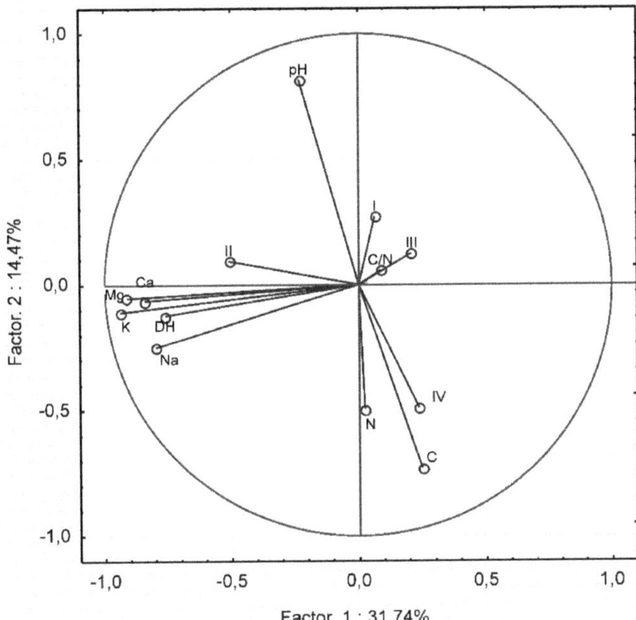

Figure 3. Projection of soil properties of birch stands chronosequence of on a plane of the first and second factors in the PCA (I—from 1 to 4 years, II—from 5 to 8 years, III—from 9 to 12 years, and IV—from 13 to 17 years; DH—dehydrogenase activity; C—carbon content; N—nitrogen content).

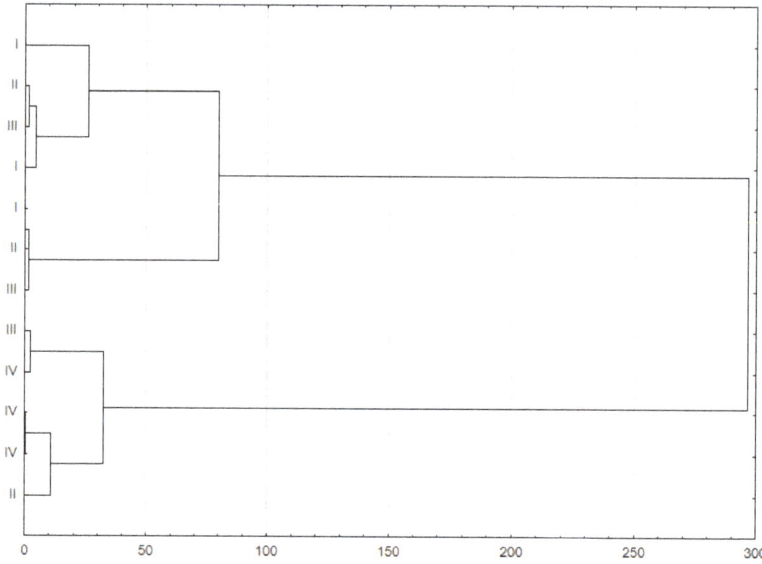

Figure 4. Dendrogram with group identified in the cluster analysis. The dehydrogenase activity and carbon content in surface layers were used for diagram preparation. I–IV—studied chronosequences of birch stands.

4. Discussion

Our results support the hypothesis that natural birch regeneration has a positive effect on the soil organic carbon accumulation. The tendency to increase carbon stocks was observed in the studied chronosequence of the birch tree stands. Several studies have shown that the decomposition of soil organic matter exceeded the input of organic matter from the trees in the initial following afforestation [26,27]. In the younger stands, the soil organic carbon accumulation was the greatest in the surface layer. In the 5-cm layer, accumulation accounted for about 30% of the total accumulation in the 50-cm deep soil column. Conversely, carbon accumulation was considerably lower than in the deeper soil layers in the 5-cm layer of the soils from the older stands. In the group I soils, accumulation in the deeper layers constituted 40% of the total stock of carbon, while it accounted for 60% in group IV. This increase in C accumulation in the deeper layers of the soils is associated not only with the processes of transporting dissolved organic compounds downwards, but also with an effect of supplying organic debris from the root systems. This is confirmed by the increase in root biomass in particular groups of stands. Subsoil soil organic carbon (SOC) storage may be promoted by the translocation of OM into deeper soil layers as DOC with the percolating water and due to bioturbation by soil animals [28]. Kotroczó et al. [29] found that plants cause greater changes in soil properties through their roots and secretions than via litter. In this sense, aboveground OM only probably has limited effects on SOM levels when compared to belowground OM [28]. Roots are a key component of the belowground part of the forest ecosystem, constituting the basic source of SOM that significantly affects soil microbiological activity [30,31]. Over time, soil organic matter input increases with the productivity of the forest stands, and the soils switch from being a C source to a C sink [32]. According to Laganiere et al. [33], the positive impact of afforestation on soil organic carbon stock is more pronounced in the cropland soils than in pastures or natural grasslands. Afforestation usually results in the establishment of higher plant biomass, and trees modify the quality and quantity of litter inputs and microclimatic conditions, such as moisture and temperature. Deng and Shangguang [34] highlight the importance of previous land use, tree species, soil depth, and forest age in determining soil C and N changes in a range of environments and land use transitions. In our study, birch stands, through aboveground and belowground biomass accumulation, had a positive effect on the quality of SOM, as expressed by the C/N ratio, which is an indicator of the extent of plant nitrogen being made available to plant residues. Li et al. [35] state that land use changes from agricultural areas to forest alter the ratios between soil C, N, and P. Springob and Krichmann [36] found that a soil C/N ratio of >20 could limit SOM mineralisation. According to Cools et al. [37], tree species are the main factor in explaining the variability of the C/N ratio. The content of better decomposed soil organic matter increases with stand age. At the same time, soil acidity and nutrient uptake increase with tree growth. Riqueiro-Rodríguez [38] note that the Pinus radiate more drastically decreases the soil pH than *Betula alba*. In another study, the acidifying affect of afforestation on mineral soil has been confirmed by a significant decrease in soil pH in the 0–5-cm layer and by a slightly weaker decrease in the 5–15-cm layer [39].

The results of the cluster analysis confirmed the distinctness in terms of C content and enzymatic activity of the soils of younger stands when compared to the soils of older stands. The soil parameters pH and soil organic carbon are important factors that shape dehydrogenase activity [15,40]. The highest pH, with the highest alkaline cation content, resulted in the highest dehydrogenase activity in soils of the younger stands (Groups I and II), reflecting the previous agricultural use of the soils and the associated intensive fertilisation and liming. According to Rousk et al. [41], pH is the main determinant of the structure of soil microbial populations. Soil pH directly determines plant growth, nutrient absorption, and the intensity of biological and chemical processes in the soil. In this work, dehydrogenase activity was positively associated with exchangeable Ca, K, and Mg contents, with a higher content of basic ions leading to an increase in pH, which results in the stimulation of soil microorganisms. Soil pH may be the major factor controlling the biomass and composition of microbial communities and their maintenance demand [42]. When assessing the properties of soils that were subjected to long-term agricultural use, several authors have considered the high plant-nutrient content as evidence of systematic fertilisation [43]. For example, Ren et al. [44] have noted that catalase, saccharase, urease, and alkaline phosphatase were significantly increased by land-use conversion from farmland to forest. According to this, significant correlations between soil enzyme activities and soil properties indicate that the soil enzyme activities are closely related to soil nutrients dynamics [18]. Dehydrogenase activity differed among the soils of the studied birch stands. The activity of this enzyme reflects that changes in the soil that are associated with the growth of the birch stands. According to previous studies, enzymatic activity is strongly stimulated by SOM [15,18], and processes that are related to organic matter transformations are carried out with the participation of soil microorganisms and their enzymes [45]. In our study, no direct relationship between dehydrogenase activity and carbon accumulation was found. Dehydrogenase activity was high in the soil of the youngest stands (first age class), and subsequently considerably decreased in class II. Our results indicate a trend to increased dehydrogenase activity in the soils of the oldest stands (IV group of stands). The highest pH, with the highest alkaline cation content, resulted in the highest dehydrogenase activity in soils of the younger stands, reflecting the previous agricultural use of the soils and the associated intensive fertilization and liming. The effects of fertilization disappear in the following years of tree stand growth. Forest stands grow and provide more litter fall to the soil, which stimulates the dehydrogenase activity. With age, greater amounts of carbon were accumulated in the surface soil layers. With increased litter input and in the absence of soil cultivation, conversion from cropland to forest could result in increased SOM stocks [46]. Similarly, Kara et al. [47] and Kang et al. [48] suggest that long-term afforestation could significantly enhance SOM contents, accumulate microbial biomass, and improve potential enzyme activities.

5. Conclusions

Our results confirm the beneficial effect of birch stand regeneration on the soil properties on post-agricultural land. We observed a clear trend of increasing carbon accumulation in the soil under the influence of birch trees. With age, greater amounts of carbon were accumulated in the surface soil layers. Dehydrogenase activity is a suitable indicator of the condition of post-agricultural soils with birch stands and, in combination with soil chemical properties, reflects historical soil management. In this sense, the determination of dehydrogenase activity allows for an assessment of the processes occurring in post-agricultural soils, which are associated with the soil organic carbon accumulation. A high nutrient content and high pH are characteristic of post-agricultural soils, facilitating a greater biochemical activity in the initial stages of stand formation.

Author Contributions: Conceptualization, T.G. and S.M.; Methodology, T.G. and E.B.; Project administration, T.G. and S.M.; Supervision, S.M.; Writing—original draft, T.G. and E.B.; Writing—review & editing, T.G.

Funding: This research was funded by Ministry of Science and Higher Education of Poland, Grant/Award Number: DS 3420/ZELiR/2018 and the Polish National Science Centre within grant N N305 400238, entitled "Ecological consequences of the silver birch (Betula pendula Roth.) secondary succession on abandoned farmlands in central Poland".

Conflicts of Interest: The authors declare no conflict of interest. The funders had no role in the design of the study; in the collection, analyses, or interpretation of data; in the writing of the manuscript, or in the decision to publish the results.

References

1. Report 2018. *Report on the Condition of Forests in Poland*; CILP: Warszawa, Poland, 2018.
2. Schoenholtz, S.H.; Van Miegroet, H.; Burger, J.A. A review of chemical and physical properties as indicators offorest soil quality: Challenges and opportunities. *For. Ecol. Manag.* **2000**, *138*, 335–356. [CrossRef]
3. Van-Camp, L.; Bujarrabal, B.; Gentile, A.R.; Jones, R.J.A.; Montanarella, L.; Olazabal, C.; Selvaradjou, S.K. *Reports of the Technical Working Groups Established under the Thematic Strategy for Soil Protection*; EUR 21319EN/3; Office for official Publications of the European Communities: Luxembourg, 2004; p. 872.
4. Steiner, C.; Teixeira, W.G.; Lehman, J.; Nehls, T.; de Macêdo, J.L.U.; Blum, W.E.M.; Zech, W. Long term effects of manure, charcoal and mineral fertilization on crop production and fertility on a highly weathered Central Amazonian upland soil. *Plant Soil* **2007**, *291*, 275–290. [CrossRef]
5. Yang, L.; Luo, P.; Wen, L.; Li, D. Soil organic carbon accumulation during post-agricultural succession in a karst area, southwest China. *Sci. Rep.* **2016**, *6*, 37118. [CrossRef] [PubMed]
6. Lal, R. Soil carbon sequestration impacts on global climate change and food security. *Science* **2004**, *304*, 1623–1627. [CrossRef] [PubMed]
7. Guo, L.; Gifford, R. Soil carbon stocks and land use change: A meta analysis. *Glob. Chang. Biol.* **2002**, *8*, 345–360. [CrossRef]
8. Korkanç, S.Y. Effects of afforestation on soil organic carbon and other soil properties. *Catena* **2014**, *123*, 62–69. [CrossRef]
9. Mao, D.M.; Min, Y.W.; Yu, L.L.; Martens, R.; Insam, H. Effect of afforestation on microbial biomass and activity in soils of tropical China. *Soil Biol. Biochem.* **1992**, *24*, 865–872.
10. Podrázský, V.; Holubík, O.; Vopravil, J.; Khel, T.; Moser, W.K.; Prknová, H. Effects of afforestation on soil structure formation in two climatic regions of the Czech Republic. *J. For. Sci.* **2015**, *61*, 225–234. [CrossRef]
11. Bolat, I.; Kara, Ö.; Sensoy, H.; Yüksel, K. Influences of Black Locust (*Robinia pseudoacacia* L.) afforestation on soil microbial biomass and activity. *iFor. Biogeosci. For.* **2017**, *9*, 171–177. [CrossRef]
12. Karlen, D.L.; Mausbach, M.J.; Doran, J.W.; Cline, R.G.; Harris, R.F.; Schuman, G.E. Soil quality: A concept, definition, and framework for evaluation. *Soil Sci. Soc. Am. J.* **1997**, *61*, 4–10. [CrossRef]
13. Gil-Sotres, F.; Trasar-Cepeda, C.; Leiros, M.C.; Seoane, S. Different approaches to evaluating soil quality using biochemical properties. *Soil Biol. Biochem.* **2005**, *37*, 877–887. [CrossRef]
14. Kucharski, J. Relacje między aktywnością enzymów a żyznością gleby. In *Drobnoustroje w Środowisku, Występowanie, Aktywność i Znaczenie*; Barabasz, W., Ed.; AR: Kraków, Poland, 1997; pp. 327–347.
15. Wolińska, A.; Stępniewska, Z.; Pytlak, A. The effect of environmental factors on total soil DNA content and dehydrogenase activity. *Arch. Biol. Sci.* **2015**, *67*, 493–501. [CrossRef]
16. Błońska, E.; Lasota, J.; Gruba, P. Enzymatic activity and stabilization of organic matter in soil with different detritus inputs. *J. Soil Sci. Plant Nutr.* **2017**, *63*, 242–247.
17. Kaczynski, P.; Lozowicka, B.; Hrynko, I.; Wolejko, E. Behaviour of mesotrione in maize and soil system and its influence on soil dehydrogenase activity. *Sci. Total Environ.* **2016**, *71*, 1079–1088. [CrossRef] [PubMed]

18. Błońska, E.; Lasota, J.; Zwydak, M.; Klamerus-Iwan, A.; Gołąb, J. Restoration of forest soil and vegetation 15 years after landslides in a lower zone of mountains in temperate climates. *Ecol. Eng.* **2016**, *97*, 503–515. [CrossRef]
19. Pająk, M.; Błońska, E.; Frąc, M.; Oszust, K. Functional diversity and microbial activity of forest soils that are heavily contaminated by lead and zinc. *Water Air Soil Pollut.* **2016**, *227*, 348. [CrossRef]
20. Józefowska, A.; Pietrzykowski, M.; Woś, B. Tree species and soil substrate effects on soil biota during early soil forming stages at afforested mine sites. *Appl. Soil Ecol.* **2016**, *102*, 70–79. [CrossRef]
21. Van der Wal, A.; van Veen, J.A.; Smant, W.; Boschker, H.; Bloem, J.; Kardol, P.; van der Putten, W.H.; de Boer, W. Fungalbiomass development in a chronosequence of land abandon-ment. *Soil Biol. Biochem* **2006**, *38*, 51–60. [CrossRef]
22. Gunina, A.; Smith, A.R.; Godbold, D.L.; Jones, D.L.; Kuzyakov, Y. Response of soil microbial community to afforestation with pure and mixed species. *Plant Soil* **2016**, *412*, 357–368. [CrossRef]
23. *WRB (World Reference Base for Soil Resource)*; FAO: Rome, Italy, 2014.
24. Ostrowska, A.; Porębska, G.; Kanafa, M. Carbon accumulation and Distribution in Profiles of Forest Soils. *Pol. J. Environ. Stud.* **2010**, *19*, 1307–1315.
25. Alef, K.; Nannipieri, P. Enzyme activities. In *Methods in Applied Soil Microbiology and Biochemistry*; Alef, K., Nannipieri, P., Eds.; Academic Press: London, UK, 1995.
26. Richter, D.D.; Markewitz, D.; Trumbore, S.E.; Wells, C.G. Rapid accumulation and turnover of soil carbon in a reestablishing forest. *Nature* **1999**, *400*, 56–58. [CrossRef]
27. Mao, R.; Zeng, D.H.; Hu, Y.L.; Li, L.J.; Yyng, D. Soil organic carbon and nitrogen stocks in an age-sequence of poplar stands planted on marginal agricultural land in Northeast China. *Plant Soil* **2010**, *332*, 277–287. [CrossRef]
28. Lorenz, K.; Lal, R. The depth distribution of soil organic carbon in relation to land use and management and the potential of carbon sequestration in subsoil horizons. *Adv. Agron.* **2005**, *88*, 35–66.
29. Kotroczó, Z.; Veres, Z.; Fekete, J.; Krakomperger, Z.; Tóth, J.A.; Lajtha, K.; Tóthmérisz, B. Soil enzyme activity in response to long-term organic matter manipulation. *Soil Biol. Biochem.* **2014**, *70*, 237–243. [CrossRef]
30. Janssens, J.A.; Sampson, D.A.; Curiel-Yuste, J.; Carrara, A.; Cenlemans, R. The carbon cost of fine root turnover in a Scots pine forest. *For. Ecol. Manag.* **2002**, *168*, 231–240. [CrossRef]
31. Kätterer, T.; Bolinder, M.A.; Andrén, O.; Kirchmann, H.; Menichetti, L. Roots contribute more to refractory soil organic matter than above-ground crop residues, as revealed by a long-term field experiment. *Agric. Ecosyst. Environ.* **2011**, *141*, 184–192. [CrossRef]
32. Holubík, O.; Podrázský, V.; Vopravil, J.; Khel, T.; Remeš, J. Effect of agricultural lands afforestation and tree species composition on the soil reaction, total organic carbon and nitrogen content in the uppermost mineral soil profile. *Soil Water Res.* **2014**, *9*, 192–200. [CrossRef]
33. Laganiere, J.; Angers, D.A.; Parc, D. Carbon ac-cumulation in agricultural soils after afforestation: A meta-analysis. *Glob. Chang. Biol.* **2010**, *16*, 439–453. [CrossRef]
34. Deng, L.; Shangguang, Z. Afforestation drives soil carbon and nitrogen changes in China. *Land Degrad. Dev.* **2016**, *21*, 151–165. [CrossRef]
35. Li, D.; Wen, L.; Zhang, W.; Yang, L.; Xiao, K.; Chen, H.; Wang, K. Afforestation effects on soil organic carbon and nitro gen pools modulated by lithology. *For. Ecol. Manag.* **2017**, *400*, 85–92. [CrossRef]
36. Springob, G.; Kirchmann, H. Bulk soil C to N ratio as a simple measure of net N mineralization from stabilized soil organic matter in sandy arable soils. *Soil Biol. Biochem.* **2003**, *35*, 629–632. [CrossRef]
37. Cools, N.; Vesterdal, L.; De Vos, B.; Vanguelova, E.; Hansen, K. Tree species is the major factor explaining C:N ratios in European forest soil. *For. Ecol. Manag.* **2014**, *311*, 3–16. [CrossRef]
38. Riqueiro-Rodríguez, A.; Mosquera-Losada, M.R.; Férnández-Núñez, E. Afforestation of agricultural land with Pinus radiata D. Don and Betula alba L. in NW Spain: Effects on soil pH, understory production and floristic diversity eleven years after establishment. *Land Degrad. Dev.* **2012**, *23*, 227–241.
39. Ritter, E.; Vesterdal, L.; Gundersen, P. Changes in soil properties after afforestation of former intensively manager soil with oak and Norway spruce. *Plant Soil* **2003**, *249*, 319–330. [CrossRef]
40. Błońska, E.; Lasota, J.; Gruba, P. Effect of temperate forest tree species on soil dehydrogenase and urease activities in relation to other properties of soil derived from loess and glaciofluvial sand. *Ecol Res* **2016**, *31*(5), 655–664. [CrossRef]

41. Rousk, J.; Bååth, E.; Brookes, P.C.; Lauber, C.L.; Lozupone, C.; Caporaso, J.G.; Knight, R.; Fierer, N. Soil bacterial and fungal communities across a pH gradient in an arable soil. *ISME J.* **2010**, *4*, 1340–1351. [CrossRef] [PubMed]
42. Chen, C.R.; Condron, L.M.; Cavis, M.R.; Sherlock, R.R. Effects of afforestation on phosphorus dynamics and biological propertiesin a New Zealand grassland soil. *Plant Soil* **2000**, *220*, 151–163. [CrossRef]
43. Wanic, T.; Błońska, E. Zastosowanie metody SIG w ocenie przydatności terenów porolnych do hodowli lasu. *Rocz. Glebozn.* **2011**, *62*, 173–181.
44. Ren, C.; Kang, D.; Wu, J.P.; Zhao, F.; Yang, G.; Han, X.; Feng, Y.; Ren, G. Temporal variation in soil enzyme activities after afforestation in the Loess Plateau, China. *Geoderma* **2016**, *282*, 103–111. [CrossRef]
45. Schimel, J.P.; Bennett, J. Nitrogen mineralization: Challenges of a changing paradigm. *Ecology* **2004**, *85*, 591–602. [CrossRef]
46. Georgiadis, P.; Vesterdal, L.; Stupak, I.; Raulund-Rasmussen, K. Accumulation of soilorganic carbon after cropland conversion to short-rotation willow and poplar. *GCB Bioenergy* **2017**, *9*, 1390–1401. [CrossRef]
47. Kara, O.; Babur, E.; Altun, L.; Seyis, M. Effects of afforestation on microbial biomass C and respiration in eroded soils of Turkey. *J. Sustain. For.* **2016**, *35*, 385–396. [CrossRef]
48. Kang, H.; Gao, H.; Yu, W.; Yi, Y.; Wang, Y.; Ning, M. Changes in soil microbial community structure and function after afforestation depend on species and age: Case study in a subtropical alluvial Island. *Sci. Total Environ.* **2018**, *625*, 1423–1432. [CrossRef] [PubMed]

MDPI
St. Alban-Anlage 66
4052 Basel
Switzerland
Tel. +41 61 683 77 34
Fax +41 61 302 89 18
www.mdpi.com

Sustainability Editorial Office
E-mail: sustainability@mdpi.com
www.mdpi.com/journal/sustainability

www.ingramcontent.com/pod-product-compliance
Lightning Source LLC
LaVergne TN
LVHW070926170726
843515LV00005B/875
9783036503288